# Working the Garden

STUDIES IN RURAL CULTURE

Jack Temple Kirby, editor

Working

# the Garden

## American Writers and the Industrialization of Agriculture

WILLIAM CONLOGUE

The University of North Carolina Press | Chapel Hill and London

Manufactured in the United States of America
Designed by April Leidig-Higgins
Set in Quadraat by Tseng Information Systems, Inc.

The paper in this book meets the guidelines for permanence and durability of the Committee on Production Guidelines for Book Longevity of the Council on Library Resources.

Portions of chapter 2 have been reprinted with permission in revised form from "Passion Transfigured: *Barren Ground* and the New Agriculture," *Mississippi Quarterly* 52, no. 1 (Winter 1998–99): 17–31, and "Managing the Farm, Educating the Farmer: *O Pioneers!* and the New Agriculture," *Great Plains Quarterly* 21, no. 1 (Winter 2001): 3–15.

Library of Congress Cataloging-in-Publication Data
Conlogue, William.
Working the garden : American writers and the industrialization of agriculture / by William Conlogue.
p. cm. (Studies in rural culture) Includes bibliographical references and index.
ISBN 0-8078-2668-5 (cloth: alk. paper)
ISBN 0-8078-4994-4 (pbk.: alk. paper)
1. American literature—20th century—History and criticism. 2. Agriculture in literature. 3. Agriculture—Economic aspects—United States—History—20th century. 4. Pastoral literature, American—History and criticism. 5. Industrialization in literature. 6. Rural conditions in literature. 7. Farm life in literature. 8. Gardens in literature. I. Title. II. Series.
PS228.A52 C66 2001 810.9'321734—dc21
2001027914

05 04 03 02 01 5 4 3 2 1

To my parents

CONTENTS

Acknowledgments ix

Introduction 3

CHAPTER ONE
Bonanza!: Origins of the New Agriculture 25

CHAPTER TWO
Challenging the Agrarian Myth:
Women's Visibility in the New Agriculture 63

CHAPTER THREE
Disciplining the Farmer:
Class and Agriculture in *The Grapes of Wrath* (1939)
and *Of Human Kindness* (1940) 95

CHAPTER FOUR
Racism and Industrial Farming:
*Actos* (1965) and *A Gathering of Old Men* (1983) 127

CHAPTER FIVE
From *A Thousand Acres* (1991) to "The Farm" (1998) 157

Postscript: Fixing Fence 185

Notes 201

Works Cited 209

Index 225

# ILLUSTRATIONS AND TABLE

## ILLUSTRATIONS

"Farming in the West—Evening" (Coffin) 30

"Ploughing" (Coffin) 30

"Harrowing" (Coffin) 31

"Sowing the Wheat" (Coffin) 31

"Reaping" (Coffin) 32

"Threshing" (Coffin) 32

"Seeding" (White) 38

"Reaping with Right-Hand Binders" (White) 39

"Shocking a Sea of Wheat" (White) 40

"Steam Threshers at Work" (White) 41

Farm plan 70

Farmall tractor advertisement, April 1939 105

Farmall tractor advertisement, January 1939 106

Photograph of *Vietnam Campesino* 139

## TABLE

Key Elements of the Competing Agricultural Paradigms 173

# ACKNOWLEDGMENTS

Many people helped to make this book happen.

Sister Christine Mihelich read the entire manuscript with care and close attention to my argument; her comments led me to rethink several ideas. Sharon Romaniello and Becky Kohinsky of Marywood University's interlibrary loan department generously went out of their way to locate books and articles that I needed. I am grateful to Marywood for a course release and a summer research grant to work on this project.

I thank Lewis Lawson for his support and example all the while I was struggling to put to paper what I was thinking. David Wyatt reminded me that the best way out is through, and he read my work with sensitivity and care. The reference librarians and interlibrary loan staff at the University of Maryland's McKeldin Library always had the answers to my frequent questions and requests—and I had many.

The advice of Jack Temple Kirby, Frieda Knobloch, and Patrick D. Murphy, all readers for the University of North Carolina Press, made this a better book. I am in debt to Sian Hunter, my supportive and helpful editor. She and project editor Paula Wald smoothed the road to publication. I thank Stevie Champion for her excellent copyediting.

My wife, Bridget, read the manuscript again and again, offering suggestions that improved its argument and style tenfold. Her patience and willingness to debate the issues that I address made this a far better work than it would have been otherwise.

Bill and Jane Conlogue, my parents, made this possible. They showed me the work and pointed the way. I dedicate this book to them.

# Working the Garden

I shan't be gone long.—You come too.—Robert Frost, "The Pasture"

**Agriculture No Longer Counts**

**IN A MILESTONE OF SORTS, U.S. TO DROP FARM RESIDENT CENSUS**

In a symbol of a massive national transformation, the federal government . . . is dropping its long-standing survey of farm residents, a striking reminder that the family farm occupies a diminished place on the American landscape. — Barbara Vobejda, *Washington Post*, 9 October 1993

# Introduction

The numbers are as staggering as they are familiar: in 1860 farmers accounted for 60 percent of the American labor force; in 1910, 30.5 percent; and by 1990, 2.3 percent (Fite, *American Farmers* 8; Howard 33). Now, in the first decade of the twenty-first century, there are too few to count. And as farmers have disappeared, so has good farmland. Between 1982 and 1992 the United States lost nearly two million hectares of cropland, an area larger than New Jersey (Gardner 6). According to the American Farmland Trust, three thousand acres of farmland disappear to urban sprawl each day (Fact Sheet). Worldwatch observers point out that 87 percent of U.S. vegetables and 86 percent of its fruit are grown in areas of rapid urban growth (Gardner 16). Meanwhile, farm herbicides contaminate the drinking water of fourteen million Americans (Lee A3), animal confinement systems pollute hundreds of rivers and streams ("EPA"; "Why the Fish"; "Rural Opposition"), and food poisoning is on the rise as more and more food is recalled by food processors ("Hamburger"). In the 1980s, while several million Americans living in poverty were malnourished, cropland lay idle, farms were sold, and distraught farmers killed bankers, neighbors, and themselves (Lamar; Ristau 16). In the 1990s mad-cow disease entered our vocabulary, along with biotechnology, bovine growth hormone, and Olestra. Archer-Daniels-Midland, the self-proclaimed "supermarket to the world," pled guilty to price-fixing and was fined $100 million (Eichenwald A1). In 1999 black farmers won a class-action racial discrimination case they brought against the U.S. Department of Agriculture ("Judge Approves"). Whereas only a few years ago farmers could see their land with their own eyes, today they use satellite global positioning systems (Feder, "For Amber" D4). American agriculture is high-tech, big business, complex—and fragile.

Is small-scale family farming the best farming? Or is industrial agribusiness? The debate has raged in the United States since the early twentieth century; its historical, economic, and political aspects have been investigated again and again. Strangely, no one has ever fully explored the literary response, an odd circumstance since literature is the place where the debate is most fully alive and where the arguments are most clearly framed. The key questions are: How do American writers understand the debate? How have they seen agriculture implicated in issues of race, gender, class, and the environment? What habits of mind do they explore—and use—to write about farming? What rhetorical and literary means do they employ to describe farming and its accelerating disappearance? This book explores answers to these questions.[1]

My analysis of a wide range of farm-centered texts illustrates that writers have been documenting America's "massive national transformation" all along. As a body of thought, these works investigate with unflinching directness how farming still feeds and reflects its cultural contexts even as it passes from the national landscape; nostalgia and pastoral assumptions find less room in writers' depictions of farm realities than they do in critics' comments about those depictions. Because the dominant urban society tends to view rural areas as pastoral retreats or as country backwaters, literary scholars assume that writers explore farming by imagining or reworking the agrarian myth, with its self-sufficient family farmer or his country cousin, the preindustrial commercial farmer. I argue that this is simply not the case. I contend that writers articulate political and social justice positions on an urban-defined agriculture whose central figure is the twentieth-century progressive farmer, the man or woman who farms according to an industrial model. In doing so, writers refute pastoral assumptions about rural life that obscure social upheavals in what many Americans too often believe is an unchanging countryside—upheavals that are national in their repercussions.

Why study key cultural issues from the perspective of the farm text? To define the debates, one must study them, literally, from the ground up. Literature is shaped by, and in turn shapes, the culture within which it evolves. Writers lay bare the language people use to perceive an otherwise chaotic reality; the best writers heighten our awareness of the metaphors we use to understand that reality and the tropes and figures we use to organize it into something we can communicate to others. Agriculture is a physical organization of the same reality—it bounds, arranges, and systematically transforms nature into something we can eat, wear, or otherwise utilize. Both are lived—and enlivening—practices that satisfy our most basic hungers. About knowing and defining the world, both do work in the world, and how each goes about its work tells us much about our soul. Just as different cultures evolve different sign systems to facilitate communication, so those same cultures create different farm systems to produce and distribute human sustenance. And, for good or ill, just as English is threatening to dominate world conversation, so American industrial agribusiness is coming to dominate world food production.

Writers aware of industrial agriculture's hegemony make connections among language, history, and farming. Jane Smiley, for example, offers this epigraph in *A Thousand Acres* (1991): "The body repeats the landscape. They are the source of each other and create each other. We were marked by the seasonal body of earth, by the terrible migrations of people, by the swift turn of a century, verging on change never before experienced on this greening planet" (from LeSueur 39). This cyclical imagery introduces a novel—modeled on *King Lear*—that probes "the loop of poison" in a farm family defined by a history of class and gender discrimination, indiscriminate violence, and incest (370). This family's story, Smiley argues, is America's story.

## Unearthing the Earth Worker in American Literature

Scholars have long read American literature through a pastoral prism. Critics generally agree that the pastoral is a literary mode marked by a protagonist's retreat into a "green world" to escape the pressures of complex urban life. In a rural or wilderness landscape, the character's interaction with the natural world restores him, and, ideally, he returns to the city better able to cope with the stresses of civilization. In a more complex pastoral, the character sees through the illusion of his turn to nature and finds himself newly aware of his fallen state in an ambiguous world. Both kinds of pastoral privilege the urban over the rural; the rural is valued only as a touchstone for the cultured, urban, therefore, civilized human. In many accounts, the pastoral mode represents the yearnings of all humankind in its nostalgic hopes for the return of a Golden Age or, at least, a simpler, happier way of life. But pastoral readings offer only an incomplete understanding of American farm literature. At once idealizing and devaluing rural life, pastoral readings divert attention from the ways in which farm literature grapples with industrial farming and the host of issues that it raises.[2]

Several of the most significant studies of American literature and culture found their arguments on versions of the pastoral impulse. For example, D. H. Lawrence claims in his groundbreaking *Studies in Classic American Literature* (1923) that American writers reflect the American need to flee civilization, "the old parenthood of Europe" (4). Building on Lawrence's work, Leslie Fiedler asserts in *Love and Death in the American Novel* (1960) that American writers seek escape in the wilderness from a feminized civilization represented by town life (14, 26). Henry Nash Smith's *Virgin Land* (1950) articulates the American "Garden of the World" myth whose central figure is the "idealized frontier farmer" (123). Smith points out that "the image of an agricultural paradise in the West, embodying group memories of an earlier, a simpler and, it was believed, a happier state of society, long survived as a force in American thought and politics" (124). In his widely influential *The Machine in the Garden* (1964) Leo Marx explores American literature through a pastoral framework, asserting that "again and again our writers have introduced the same overtones, depicting the machine as invading the peace of an enclosed space, a world set apart, or an area somehow made to evoke a feeling of encircled felicity" (29). Rereading the pastoral from a feminist perspective in *Lay of the Land* (1975), Annette Kolodny argues that at the center of a "uniquely American pastoral vocabulary . . .

lay a yearning to know and to respond to the landscape as feminine" (8). In *The Environmental Imagination* (1995) Lawrence Buell locates his argument within this tradition, noting that scholars have long had a "critical urge" to explain American literature and culture in terms of the pastoral mode (33).

But when the pastoral is brought to bear on farm texts, critics sometimes too easily leap from farmer to shepherd, from farm to garden. A case in point is Willa Cather's *O Pioneers!* Attention to its pastoral elements dominates scholarly attention to the novel. Forgetting for a moment the actual work Alexandra does to create her farms, critics often imagine her connection with the land in idyllic terms. John Murphy, for instance, asserts: "Like the generating light God made to govern the day, the sympathy evident in Alexandra's radiant face has made the wild land prolific" (116–17). John H. Randall argues that *O Pioneers!* portrays "the Garden of Eden . . . as existing in that short interval of time between the passing of the pioneers and the completion of settlement by the farmers" (74–75). Similarly, Susan Rosowski claims that the novel's narrator "imaginatively transforms Nebraska into a New World Eden" (*Voyage* 144). Though nodding to how Alexandra does business, Janis Stout points out that her "farming methods nevertheless draw on a 'feminine' spirit of cooperation with the land and intuitive understanding of her animals' needs, quite different from her father's method of trying to master or 'tame' his 'wild land' " (102). Most recently, Mary Paniccia Carden explores how *O Pioneers!* supplements the "self-made man" in the American romance of nation-building with a self-made woman who "understands the prairie as a growing and vital entity unto itself and honors what it is naturally inclined to grow rather than forcing incongruent production on it" ("Creative Fertility" 279, 281). These are valuable and often brilliant readings, but I believe that there is more to the story. As I argue in chapter 2, the novel is less about pastoral unions with the land than it is about celebrating industrial farming.

Scholars who examine the pastoral impulse in American literature give us much to think about, but the critical preoccupation with the pastoral obscures the possibility of other readings. Whereas many scholars nod to Virgil's *Eclogues* in reading American literature, I root my analysis in Virgil's *Georgics*, a four-book poem that celebrates farmwork. Best read, like the *Eclogues*, as a text that represents a particular vision of human life, the poem occupies a position midway between the heroic ideals of Virgil's martial *Aeneid* and the happy leisure of his pastoral *Eclogues*. The *Georgics* outlines a mode of thought necessary to sustain human life: hard work is inevitable and creative (Book 1, lines 121–36), variety should rule (Book 2, lines 84–109), human life is communal (Book 1, lines 300–305), humans ought to heed nature's patterns (Book 1, lines 50–53). An "organic patterned poem," the *Georgics*' "whole moral fabric" rests on the "old-fashioned yeoman" who "must work for himself" (Wilkinson 106, 54).[3]

Whereas the pastoral mode stresses retreat and return and the fantasy of living in full communion with nature, the georgic explores the lived landscapes of rural experience. It is here that our ambiguous and contradictory relationships with nature are most obvious. This may be why, in depicting as it does the "complexity of the local" in everyday life, the *Georgics* offers "for the first time extended natural description" (Heinzelman 188, 205). Suggesting the complicated ties we all have to the natural world, the poem represents farmers as "sustaining" and "helping" but also as "aggressive and destructive" (Perkell 37–38). "Marked by a sense of limits," the georgic mode understands history as "embedded, repetitive, and inescapable" (Feingold 69; Heinzelman 184). But pastoral nostalgia finds little room in the georgic: "For nostalgia for a past Golden Age, Virgil substitutes the vision of cooperative effort to bring about a Golden Age in the future" (Low 138–39; see Perkell 20). To represent the creation of a better world, the georgic emphasizes "the performance of equitable labor for the common welfare" (Low 18). Its yoking of justice and hard work prompts an important question: "How does [work] remain virtuous and under what conditions do sharpened human wits continue in their devotion to civilizing toil rather than aggressive greed?" (Wilkinson 59; Feingold 24). Wrestling with this question, the poem leads the reader to an "identification with the experience of the other and thus generates a moral community"—one that understands the "vulnerability and weakness of all human enterprise" (Perkell 20). In imagining varieties of work, the georgic argues for the transcendence of "social disparities," working with and within nature, and for diversity—of plants, animals, and humans (Perkell 20).

An intersection of intellectual/abstract and bodily/concrete ways of knowing, the georgic vision is especially significant today, a time when we are becoming increasingly aware of how we are changing the landscapes we live within and depend upon—and how those changes change us. Literally, georgic means earth worker, farmer. To unearth the earth worker means to realize that we are all, whether we like it or not, working the earth: We all eat, and through us passes the work of others. To unearth the earth worker in American literature means to realize that our national literature has long confronted basic georgic questions: How have we been at work in the world? How have we enacted on the ground what we think in our heads? What world do we want to create or maintain? How can we do it responsibly?[4]

I offer my argument as a contribution to contemporary ecocriticism. Until recently, literary criticism was primarily concerned with literary depictions of human culture and not nature, with literary varieties of human social relations and not our ties to the natural world. Since the mid-1970s, a growing body of scholars has been studying in earnest "the relationship between literature and the physical environment" (Glotfelty xviii). Unfortunately, discussions of this connection focus overwhelmingly on wild nature, not on farm landscapes. As a recent example, the board of directors of the Association for the Study of Literature and the Environment was initially lukewarm about holding the June 2000 Food and Farming in American Life and Letters Symposium (Wallace). Did the board fail to see a connection between nature and farming? Or is the wild poetic and farmland merely prosaic? If we do not pay close attention to how rural landscapes are imagined, what chance do we have of renewing our living ties to the natural world? To ignore this ground is to pave over the complexity of the human/nature mosaic.

Though many critics today seek to redefine the pastoral, I suggest that stopping there will only undermine efforts to develop a truly complex ecocritical reading of American literature (Love 234–35; Buell 52; L. Marx, "Pastoralism" 66). The argument that humans are within nature and not outside it will fail unless more ecocritics accept that humans are in a unique position: "We are, whether we like it or not, the lords of creation; true humility consists not in pretending that we aren't" (F. Turner 50). Like many environmentalists, some ecocritics are reluctant to ask georgic questions, I believe, because questions about the uses of nature imply for them complicity with forces that seek to erase or dominate the natural world (R. White 172). I disagree. Georgic readings examine—in all its forms and consequences—our work record, the history of the intersections we have made among human work, human imagination, and the physical environment. Our current arguments about American literature do not adequately address how questions about work, community, technological restraint, and human uses of nature changed with the introduction of an urban-defined industrial agriculture that has erased the pastoral's central tension between city and country. Rather than wishing it away, American writers have been wrestling with industrial agriculture and its assumptions about places, animals, plants, and people, assumptions that are running rural landscapes to ruin.

The georgic is especially important now because the pastoral posits a distinction between rural and urban that no longer holds. The redefinition of farming on an urban industrial model—begun in the late nineteenth century—is nearly complete; the country and city now possess a "shared consciousness" (R. Williams 295). In less than a hundred and fifty years, we have removed ourselves from a basic awareness of how we maintain ourselves on the planet; we have wiped from our cultural memory an awareness of what it means to work with the land. Separated from our sources of sustenance, we no longer understand that our lives depend on a few inches of topsoil and some regular rainfall. This disconnection is a large part of everyone's worldview, even those who should know better. What else explains the spectacle of farmers on food stamps during the 1980s Farm Crisis?

My work is significant, I believe, because it ties the nation's most life-sustaining activity—food production—to how it thinks through its most pressing and potentially explosive issues: widening environmental degradation, escalating racial tensions, intensifying class divisions, and increasing pressures on maintaining safe food and water supplies. With food production now concentrating in fewer and fewer hands, we should not be surprised if these issues fray social and ecological fabrics even more in the coming century as more and more people worldwide scramble for less and less available and higher-priced food. Efforts to sustain ecologically friendly ways to live and farm will be damaged from the start unless we can imagine new ways to see how writers have been concerned with food and farming all along. The georgic asserts that the human heart beats in tune with the rest of creation; it offers the seed of a new way to read.

My argument, then, is a georgic one. I believe that American writers are far more interested in debating how best to work and transform the American "garden" they find themselves in than they are in simply contemplating it or wishing for an ideal state of innocence. The pastoral mode is concerned too much with the contemplation of nature rather than with its transformation, with leisure rather than work, with the past rather than the present and future, with idyllic stasis rather than change. I contend that writers ask how best to use (or not use) the machine in the garden, how the garden and its people are affected by the adoption of that machine, and how much it might cost us, as human beings and as members of the biosphere, to deploy that machine. The best way to study these complex issues is by examining the ground where they are most carefully connected and most fully imagined—the literary farm text.

## Nineteenth-Century Agrarianism

The dominant definition of American agriculture through the nineteenth century was a "literary agrarianism derived originally from classic antiquity" (Johnstone 116). An eighteenth-century intellectual and political construction, this literary agrarianism was decidedly pastoral (L. Marx, *Machine* 126; Hofstadter 25). Concerned with actual farmwork only tangentially, its principles prescribe that farmers—and they are always male—are virtuous, hard-working, independent, happy, neighborly, family-sustaining, and faithful to the republic and to God, all because they work closely with nature. Standing behind this was the Old World hope that the New World was a second chance, a new Garden of Eden.[5]

Two important eighteenth-century intellectual ideas that influenced American agrarianism were "the theoretical teaching of the French Physiocrats that agriculture is the primary source of all wealth, [and] the growing tendency of radical writers . . . to make the farmer a republican symbol instead of depicting him in pastoral terms as a peasant virtuously content with his humble status in a stratified society" (H. N. Smith 128–29). In American agrarianism, not only is the farmer symbolic of the new republic but also he is morally superior to his urban countrymen. Thomas Jefferson made this claim explicit in his *Notes on the State of Virginia* (1787): "Those who labour in the earth are the chosen people of God, if ever he had a chosen people, whose breasts he has made his peculiar deposit for substantial and genuine virtue" (217; see Hofstadter 27–29). Jefferson's yeoman, a self-sufficient man free from "the casualties and caprice of customers," owned a small piece of land that he and his family farmed to provide for themselves the essentials of life—food, clothing, shelter—plus surplus commodities for export to the "mobs of great cities" in Europe in exchange for manufactured goods (*Notes* 217). This ideal farmer, who was to be the backbone of the United States, was a man interested mainly in preserving his family's presence on the land through several generations. Working hard within the compass of the natural world and feeling a spiritual as well as a physical kinship to nature, the yeoman held dear those virtues that the new nation would rely on for its prosperity: frugalness, hard work, charity toward others, and love of God and country. Politically motivated, Jefferson's espousal of the agrarian myth was a defense of a national economy based on an agricultural model whose foundation was an independent husbandry that would never be at the mercy of capitalist businessmen or entrepreneurial landlords.

Jefferson later rethought this definition of the ideal farmer (Letter 549; Baum 12), but the myth he articulated in his *Notes* persisted into the next century because it flattered farmers, who were the majority of the nation's population, and because it perpetuated an ancient vision of rural life that extends at least as far back as Virgil (Hofstadter 29–30). One legacy Jefferson left the nineteenth century was that farming, and more generally, country life, was to be represented in politics and in literature according to his agrarian ideal. For example, advocates of the 1862 Homestead Act "sincerely believed that the yeoman depicted in the myth of the garden was an accurate representation of the common man of the Northwest, and this belief was evidently shared by thousands of voters" (H. N. Smith 172). As the century wore on, however, the agrarian myth retreated to the minds of those who guided national thinking about farming, many of whom were not farmers (Fite, *American Farmers* 10). Literary scholars who write about farm texts today may be the final, innocent defenders of this agrarian ideal.

## The Old Agriculture

Despite late-nineteenth-century political and literary rhetoric, farmers did not simply accept the agrarian myth: "Judging from the evidence provided by the agricultural press, not only did farmers fail to believe the agrarian myth; they were not even sure what the agrarian myth was" (Abbott 48). By 1870 at least, many farmers understood themselves to be businessmen rather than Jeffersonian yeomen (Rome 38). But, though they were commercialists before 1900, the vast majority of farmers were preindustrial when the new century dawned (Danbom 12). In advocating the industrialization of the farm, early-twentieth-century agricultural reformers dismissed the preindustrial farmer as the "old farmer" (Adams 11).

Typical nineteenth-century farmers were independent and to some degree self-sufficient (Adams 14; Fite, *American Farmers* 3). Living mainly on small, diversified farms, preindustrial farmers grew a variety of crops and raised several kinds of animals to supply family needs first. A "jack-of-all-trades," many farmers spent their days working with handheld tools, such as hoes, scythes, and broadcast seeders (Adams 11). Though after the Civil War horse-drawn machinery was increasingly used, most farmers had few machines and what they did have they kept for as long as possible and usually repaired themselves. As late as the 1930s, "the three-mile-an-hour gait of the horse established the speed and power of most field work on American farms" (Fite, *American Farmers* 66).

The preindustrial farm community exchanged machinery, labor, and knowledge in both noncommercial and commercial ways. The barter system dominated many places and times; farmers exchanged foodstuffs for manufactured goods and often swapped labor and machinery with neighbors. Specialty crops sold to consumers in towns and cities depended on the availability of transportation, but as transportation improved and as more laborsaving machinery was introduced, regional specialization became common. Though farm size grew, the number of acres generally remained manageable by a single farm family.

Farm labor was done primarily by family members. Husbands were usually decision makers and the principal fieldworkers. Children did chores appropriate to their ages. Wives worked mainly in the home, though they were ready to do fieldwork when necessary. Hired help was often someone familiar with the family and with farming and lived in the farmer's home. Under constant threat from weather changes, work hazards, and market prices, farm life was difficult. Work hours were long; the work itself was often monotonous and physically demanding (Fite, *American Farmers* 11). Farming was no life of repose; nature was an exacting taskmaster. This garden did not take care of itself.

Many contemporary observers believed that the late nineteenth century's accelerating exodus of rural youth to cities was a direct result of a lack of intellectual stimulation on farms (Fite, *American Farmers* 12). Because of relatively poor transportation and communication networks, most nineteenth-century farmers were isolated from the amenities of urban life and "suffered from low social status" as a result (Fite, *American Farmers* 11). Many Americans viewed farming as something that anyone could do and defined the successful rural youth as someone who left the farm to take up a profession in law, medicine, or government.

Suspicious of "book farming," the preindustrial farmer relied on inherited farm knowledge (Adams 69). Farmers often dismissed the advice of scientists and other experts who worked in the agricultural colleges established during and after the 1850s. Well into the twentieth century many farmers continued to plant and harvest by the phases of the moon, and few kept track of their farms in any systematic, businesslike manner. Looking back in 1897, the *Cultivator and Country Gentleman* noted that "such accounts as were kept were as often scrawled on the barn door as entered in a more formal ledger" ("Business"). Most farmers remained fiercely independent in attitude, even as they lobbied the federal government for help—an increasing occurrence in the late nineteenth century.

As the century ended, more people were calling attention to a paradigm shift within agriculture, from understanding farming as a way of life to understanding it as a business. In November 1897 the editor of the *Cultivator and Country Gentleman* took note that the urban *Scribner's Magazine* had recently published contrasting visions of farming in back-to-back articles. William Allen White's piece, "The Business of a Wheat Farm," part seven of a series entitled "The Conduct of Great Businesses," reports on great wheat farms in North Dakota; Walter A. Wyckoff's "A Farm-Hand," the fourth installment in his series "The Workers," describes a typical nineteenth-century preindustrial farm in Pennsylvania. Promoting White's definition of agriculture, the *Cultivator and Country Gentleman* editor points out to his farm readers: "One who has left farming alone, or only dabbled in it, is apt to associate the occupation with ideas of leisurely activity of an old-world sort, rather than with the ceaseless tumult and bustle, the whirr of machinery, the minute and painstaking calculations and book-keeping that characterize business enterprise" ("Business"). He then quotes parts of White's article to illustrate North Dakota farms' huge size, their costs and profits, farm laborers, and machinery—descriptions that were worlds away from the agriculture depicted in Wyckoff's piece and from the experience of most American farmers at the time.

William Allen White describes an industrial agriculture: the "farmer of to-day" is a "capitalist, cautious and crafty . . . an operator of industrial affairs, daring and resourceful" (531). He reports that owners of the great wheat farms do not work the land themselves; they hire managers and skilled, all-male work crews (538). Laborers and managers work in rigidly defined social hierarchies in which they are housed and fed separately (538–40). A typical Dakota wheat farm has three divisions of several thousand acres, each managed by a superintendent who purchases trainloads of machinery (538, 545). On such farms, farming is big business.

Wyckoff portrays a preindustrial farm. Working his way west as an unskilled laborer, Wyckoff was hired in September 1891 by a Mr. Hill, "one of the best farmers of the neighborhood" (554). Hill's family farm, near Williamsport, Pennsylvania, is self-sufficient in its diversity. Hill grows corn and apples and raises cows and pigs and has no hired man until Wyckoff. That Hill keeps his plows and "other farming tools" under cover suggests to Wyckoff "the active, thrifty strength of wise economy" (554). Work on the Hill farm is divided by gender: Hill's wife and daughter work at the house while Hill is in the fields. But the distinction is not absolute: the women work with Hill in the barn at milking time. In a typical act of preindustrial farm cooperation, Hill and his brother, who owns a neighboring farm, work together to rebuild a pond dam.

Hired man and employer are social equals on this farm. While working for Hill, Wyckoff is not conscious of a "boss [who] stands guard over me as a dishonest workman" (555). When Hill heartily approves of Wyckoff's work on the dam, Wyckoff feels a "sense of proprietorship" that compensates as much for his hard work as his pay (556). The men then work side by side picking apples, which Wyckoff enjoys as "Arcadian in its joyous simplicity" (556). As they work, they become "chummy"; they discuss a range of topics—everything from English politics to Roman architecture (556–57). Only because there is no room in the house, Wyckoff sleeps in the barn, but he eats his meals at the family's table (555–56). After supper on Sunday, Hill and Wyckoff relax and converse in the front yard (558). On farms like this one, farming is a way of life.

## The New Agriculture

Though always attuned to new scientific and technological advances, farmers confronted a new phenomenon in the late nineteenth century: an urban re-creation of farming on an industrial model. The farmer's redefinition was imagined and promoted by people removed from agriculture, just as Jefferson's agrarianism had been. But now professional experts rather than statesmen or poets defined farming; to feed and provide labor for the late-nineteenth-century industrial economy, experts convinced farmers to industrialize. Their ring of authority was unmistakable. Couching their beliefs in rationally discovered principles, these reformers argued that the industrial farmer was the final step in a natural evolution of the agriculturalist from pioneer to manufacturer.

Farm industrialization is not simply synonymous with the use of machinery or scientific methods; preindustrial farmers used machinery—horse-drawn seeders, for example—and scientific methods such as fertilizers and selective breeding. Agricultural industrialization requires farmers to conceive of plants, animals, land, and people through a narrow mechanistic frame that tends not to see them as living things. The industrial farm works toward ever-greater control over nature as a factor in production rather than working with it. Profit is the measure of the new farm, not a family's continuance on the land, its quality of life, or its relations to the larger community. The new farmer rejects traditional conceptions of agricultural work, work whose model is the husbandman.

The new farmer disdains the conflation of management and labor in the figure of the farmer, the privileging of inherited farm practices, the recognition of immanent value in work and property as opposed to their exchange value, the noncommercial networks of exchange within a community. In contrast to traditional notions of farmers as husbandmen, industrial agriculture looks to business and science as models. Its basic precepts include division of labor, adoption of the latest methods and machinery, systematic business management and book farming, heavy participation in a cash market that leads to specialization, emphasis on change and experimentation, and reliance on experts outside the community for reliable advice. Industrial agriculture aggressively seeks to replace haphazard tradition with rationality, systematization, efficiency, organization, professionalization, and an identification of farming with urban manufacturing.

By the 1880s and 1890s the agricultural press was representing farmers, not apart from manufacturers, but as manufacturers—the measure of farm success would no more be the well-kept homestead; it was to be the most efficient, most profitable business in a new industrial order. Hand in hand with this redefinition was the pastoral cast it was given to make it more palatable to the nation: "among some professional agricultural leaders and educators there has evidently been a desire to idealize rural life in a moral and aesthetic way, and also to see agriculture principally in terms of the most prosperous group of farmers" (Johnstone 165). That these leaders and educators were usually professionals with tenuous connections to farming, that they were usually urban, white males with ties to powerful business interests who promoted primarily the economic interests of the wealthiest farmers is significant when one remembers that their work most often negatively affected small farmers, women, and minorities.

Efforts to redefine the farmer in industrial terms began at least as early as *Farm Journal*'s 1890 proclamation: "We farmers are manufacturers, and when we adopt the successful manufacturers' emphatic methods we shall succeed as well as they" ("We, Too"). The *Journal* emphasized abandoning old methods for the "newest and best. . . . Hard thought must evolve new plans," and "shorter, cheaper methods must be made to supersede the older." In 1907 Kenyon Butterfield, the father of rural sociology (Danbom 43), urged Americans to "eliminate" the farmer who "is dazzled by the romantic halo of the good old times" and to replace him with the "new farmer," who is characterized by "keenness, business instinct, readiness to adopt new methods. . . . He is a successful American citizen who grows corn instead of making steel rails" (65, 55, 57).

Redefining the farmer became a "national issue" in 1908, when President Theodore Roosevelt formed the Country Life Commission to study "the problem of farm life" (Neth 98). The commission defined "two great classes of farmers: those who make farming a real and active constructive business, as much as the successful manufacturer or merchant makes his effort a business; and those who merely passively live on the land" (*Report* 85). In contrast to those who "refused to become modern," the new farmer's "business [was] gradually assuming the form of other capitalized industries" (Neth 98; Davenport 49).

The Country Life movement was the rural manifestation of the national Progressive movement (Fink 24–25). Middle-class urban intellectuals with rural roots, most Country Life leaders were educators and journalists, many were involved in the Conservation movement, and several published works advocating agricultural and educational efficiency (Bowers 31–33). Urban agrarians, a vocal subset of Country Lifers, were social thinkers who looked "to the countryside for solutions to urban problems . . . for correctives to urban values. For them, rural America symbolized what America had been and was an antidote for what it was becoming" (Danbom 25). Concerned with rural uplift and uneasy about the nation's burgeoning industrial system, these thinkers saw the farmer as a "hard-working small capitalist" whose role was to be a "'harmonizer between capital and labor'" (Danbom 27). Country Lifers firmly believed that the countryside was to supply cities with its best people; in their view, urban leaders ought to be rural men (Bowers 36).[6]

The New Agriculture demanded a new agrarianism that could accommodate Jefferson's yeoman and the new farmer. The drift toward altering the myth can be seen in 1906, when William Sumner Harwood conflates old and new definitions of farming in ending *The New Earth*. He praises industrial agriculture in terms derived from the agrarian myth:

> And the farm, the new farm, with its free life, its breath of the open, its close touch with nature, its hard but never menial labor, its refined home life, its articulation with all that is best in modern life, this mighty manufacturing plant of the New Earth, is turning out not only the unthinkably valuable products and steadily heaping up billions upon billions in its reserve, and maintaining at a high plane the very life of the race, but it is manufacturing men and women,—sane, symmetrical, stocked with common sense, open to higher things, receptive and retentive, untainted by speculation, and bearing a bitter hatred of the greed that not permanently, but with infinite disgrace, has fastened itself upon America. (378)

The next year Kenyon Butterfield, then president of the Massachusetts Agricultural College, took another step toward a new agrarianism when he drew distinctions among three types of farmers: "the new farmer," "the mossback," and "the old farmer" (54–55). His description of this triad not only reveals how proponents of the New Agriculture convinced people that industrial farming was an improvement on older farm practices but also illustrates how the agrarian myth could be used to persuade them. Butterfield defines farmers who refuse to adopt new practices as mossbacks: "The mossback is the man who has either misread the signs of the times, or who has not possessed the speed demanded in the two-minute class" (56). For Butterfield, the "contrast is not between the old farmer and the new, for that is merely a question of relative conditions in different epochs of time. The contrast is between the new farmer and the mossback, for that is a question of men and of their relative efficiency as members of the industrial order" (57). His "old farmer" is the yeoman of yesteryear who was a new farmer in his day. In expressing his argument this way, Butterfield redefines the true inheritor of Jefferson's agrarianism as a man who changes with the times. Thus the independent, self-sufficient farmer could live on in the nation's mind, even as tenancy rates rose and cities filled with displaced farm people: "it is a fact that this idyllic agrarian fundamentalism has been perpetuated principally by the intellectual and reform elements that have been most active in modernizing American agriculture" (Johnstone 165). By simply glancing at farm publication advertising, one sees that Butterfield's agrarianism remains with us.

## Defining the Farm Novel

Few scholars have bothered to define the farm novel, though many agree that "the development of American rural fiction has taken place within the twentieth century . . . rural novels published during the first decade of the century were so few in number and usually so lacking in strength that the real development of farm-life fiction may be said to have had its beginning about 1910" (Sherman 67). The farm novel "which treats farm life seriously, realistically, and as the main subject is largely a phenomenon of the twentieth century" (Meyer 13); according to John T. Flanagan, "As late as 1900 only a handful of genuine farm novels had appeared" (113). The recognition of a rural problem in the earliest years of the century, the fear that the countryside was being depopulated, and the consequent creation of a nostalgic urban audience may account for the creation of the farm novel as a distinct genre only after 60 percent of the nation's population had been absorbed by American cities.

Roy Meyer does offer a definition of the farm novel:

> 1. "it deal[s] with farm life"—the plot is farm-related, characters are farmers, the setting is a farm;
> 2. displays an "accurate handling of the physical details of farm life";
> 3. uses the vernacular;
> 4. reflects "attitudes, beliefs, or habits of mind often associated with farm people. . . conservatism, individualism, anti-intellectualism, hostility to the town, and a type of primitivism," and an emphasis on hard work and the supposed unity of man and nature, both spiritually and physically, that makes farming "intrinsically more wholesome" in contrast to the corruption of city life. (7–12)

Meyer's first two points are obvious enough, and point three could as easily define other genres—for example, urban neighborhood novels. But point four's habits of mind are attitudes that urban readers expect of farmers and are rooted in the beliefs of early-twentieth-century urban agrarians and social scientists. The new farmer is not fiscally conservative; he or she joins cooperatives or answers to stockholders, appeals to experts for advice, enjoys urban comforts and adopts urban attitudes, and sees farming as a business, not a way of life.

In addition to Meyer's first two characteristics, I suggest the following as key components of a farm novel. I limit this list to balance rigor with flexibility. No text will satisfy each stipulation, of course, and these components are not intended as the final word:

1. *A farm novel explores interrelationships between human work and nature.* In seeking to understand this relationship, the farm novel foregrounds the local and the particular; its landscapes are not simply settings for human action but places where the human, animal, plant, and nonorganic worlds interact in interdependent relationships. Characters are not passive in the face of the world they confront; they are active participants in creating or changing the world around them. They are keenly aware that work has consequences.
2. *A farm novel confronts history.* Every farm is the accumulation of past acts and embodies a series of reciprocal relationships between the human and the natural, good, bad, and indifferent. Aware that past human work has demeaned or enriched places and people, the farm novel explores how local, regional, and national history has brought people and places to what they are. Because urban industrial farming tries to erase history by making each farm like every other—one factorylike chicken farm is pretty much like the next one—texts concerned with industrial farming subject its practices to historical analysis.
3. *The farm novel investigates the health and wholeness of farm families and communities.* A farm may be defined as a web of people, plants, and animals occupying a particular place. A family-run mixed farming enterprise is the most obvious example. But so, too, is the corporate farm raising several thousand hogs on a few acres. Which is the healthiest—and in every sense: ecological, economic, moral, ethical? What does the health/wholeness of farm communities say about the health/wholeness of the larger community?
4. *The farm novel investigates the impact of technology on people and places.* A farm is also a collection of technologies, and the farm novel maps the scale and costs of using a technology—the tractor, for instance. Frequently

linking natural disaster to human technological hubris, the farm novel also explores how industrialization changes the web of relationships among people, within places, and between people and places. A central change among humans is that while his neighbors streamed to the city, the twentieth-century industrial farmer found himself there without leaving home.

Writers make new the stories we live by. A living expression of human perception, literature shapes our reality by revealing to us how we see each other—and the natural world. Enlarging our capacity for making fruitful and lasting connections, literary works suggest how we imagine ourselves: They are spaces where we remember the pieces of our lives by seeing ourselves reflected in the lives of others. After all, to write is to learn; to read is to connect. To remember others is to define the people and the world around us as more than human and natural resources. If we are not mindful of our neighbors, how can we ever truly know the world as home and other life as family members?

## An Outline

This book examines selected farm texts within the context of late-nineteenth- and twentieth-century agricultural history. So that my analysis moves across space as well as time, I have chosen works from a variety of genres written by authors from different farm regions. I show that from the late nineteenth century through the 1920s, many writers supported industrial farming, but that with the Great Depression, the Dust Bowl, and the loss of thousands of small farms in the 1930s they began questioning its advantages.

I organize my argument within a race/class/gender framework because to talk about American agriculture is to remember the erasure of Native American populations, the enslavement of Africans, the displacement of rural poor to urban factories, the exploitation of migrant workers, the silence and invisibility of women. White, upper-middle-class men have always been the principle leaders and defenders of industrial farming. Their usual challengers have been farm laborers and migrants, generally people of color, women, and the poor. The social, economic, and political confrontations between these groups have shaped and propelled the rancorous debates over the proper way to feed us all. Most recently, however, a growing cross-section of Americans has joined the fray by questioning industrial farm practices and calling for alternatives in the form of community-sustained agriculture and organically grown food.

But the numbers are stark: in 1997 there were 1,911,859 farms in the United States. Of these, only 165,102 were operated by women, 27,717 by Hispanics, 18,451 by blacks, 18,495 by Native Americans, and 8,731 by Asian Americans (*Quick Facts*). This totals 238,496 farms; white men presumably owned the other 1.6 million. In 1997 the largest farms—those over 2,000 acres—constituted 4 percent of all farms; those with over a million dollars in sales made up 1 percent of farms and accounted for 42 percent of all farm sales (*Quick Facts*). Is this a democratic distribution of the sources of life?

Hardly inheritors of Jefferson's agrarianism, owners of today's largest farms are wealthy. For example, in 1994 the *Washington Post* reported that several California farmers owed millions in low-interest loans to the federal Farmers Home Administration, a Depression-era program created to help struggling small farmers. According to the *Post*'s investigation, one cotton/alfalfa farmer, a collector of vintage military aircraft, owed $1.2 million (LaFraniere A16). Another farmer, a resort owner, was "just below the Forbes magazine cutoff for America's 400 richest families"; he reportedly owed $17.6 million in loans (LaFraniere A16). The political voice of "farmers" like these is the billion-dollar, tax-exempt American Farm Bureau, profiled in an April 2000 *60 Minutes* investigation as opposed to minimum-wage laws, prohibitions against offshore oil drilling, the Equal Rights Amendment, and the 1965 Voting Rights Act ("Voice" 19, 24–25).[7] This is the green world of modern farming.

To define the industrial farm, I begin by describing the rhetorical origins of industrial farming in the United States as beginning around 1880, earlier than many might think. Urban journalists writing at the time gave the nation the language that it needed to conceive the new farm. Announcing to their urban audiences the appearance of bonanza farms in North Dakota, these journalists argued that farming was a business on par with urban industry. Their work pointed the way to Frank Norris's *The Octopus* (1901), a novel that appeared just as urban progressives began clamoring for a redefinition of farming. Norris's work illustrates how industrial farming defines nature as an abstraction.

Alexandra Bergson in Willa Cather's *O Pioneers!* (1913) and Dorinda Oakley in Ellen Glasgow's *Barren Ground* (1925) are women who choose the techniques of industrial agriculture to free themselves from male dominance. These novels expose the gender assumptions of an American culture whose understanding of farming is rooted in an agrarian myth that defines farmers exclusively as men. Though Alexandra and Dorinda refute the image of farm women as homebound farmwives, their acceptance of industrial agriculture acknowledges its assumptions about hierarchies of labor, control of nature, and class distinctions. These novels appeared at important moments in farm history: Cather's novel was published in the golden age of American agriculture just prior to World War I; Glasgow's appeared a dozen years later in the midst of a farm depression that foreshadowed the Great Depression of the 1930s.

Industrial farming rigidly stratified the relatively fluid class structures within preindustrial farming, a fact represented in Ruth Comfort Mitchell's *Of Human Kindness* (1940), a novel written in response to John Steinbeck's *The Grapes of Wrath* (1939). Both novels depict how industrial farm practices affect class relations among people whose lives are touched by them. Each appears near the end of the Great Depression, and both make significant statements about the history and causes of that economic and social catastrophe. Read together, they illustrate the nation's conflicted understanding of farm realities. Whereas Steinbeck hammers away at the industrial farmer in his portrait of dislocation, joblessness, and violence, Mitchell argues in favor of the industrial farm ideal in her portrait of rugged individualism, patriotism, and rural "progress."

Race played a key part in agriculture's industrialization. Historically, whites have owned most land and machinery, and blacks, Hispanics, and Asians have been laborers. Chicano playwright Luis Valdez's *actos* focus on ties between racism and corporate farming in California's San Joaquin Valley. Valdez's actos were a new, Chicano theatrical form born in the Cesar Chavez–led grape strike of 1965. Chavez's work called national and lasting attention to the plight of migrant farmworkers.

The U.S. Commission on Civil Rights first reported in 1982 that the African American farmer was threatened with extinction. Suing the U.S. Department of Agriculture in 1997, black farmers claimed that institutionalized racism was one reason for their dwindling numbers ("Judge Approves"). In 1983 Ernest Gaines published *A Gathering of Old Men*, a novel examining the lingering effects that industrial farming has had on relations among African Americans, Cajuns, and plantation-owning whites in a rural South bearing the burden of a history of slavery. Gaines's work appears, too, at the onset of the most severe post-1930s farm depression, one that drove thousands more black farmers from the land.

Set in the same time period as Gaines's novel, Jane Smiley's *A Thousand Acres* (1991) traces the human costs of industrial farming's late-twentieth-century triumph. But even as this novel was being written, rural sociologists were noting an emerging paradigm shift within farming that might return the nation to a less industrial agriculture that values family, community, and land as interdependent, living entities. A key proponent of this alternative agriculture is poet, novelist, essayist, and small farmer Wendell Berry. Together, Smiley and Berry point to the divide separating those arguing over agriculture's future in the new millennium.

Recognizing how writers represent farmers does not merely change our perceptions of farm texts—it profoundly alters how we conceive American literature. Seeing canonical texts through a more georgic prism reveals how closely American farm literature follows the contours of American farm history and politics. In this light, American literature becomes not a constant tension between what Americans have and the mythic place of repose they want but an ongoing debate about how best to work the American "garden."

**The farmer is the true aristocrat here—the man of means and elegant leisure.**
**—Hiram Drache, *The Day of the Bonanza*, 1964**

CHAPTER ONE

# Bonanza!

## Origins of the New Agriculture

Working his fields, a farmer comes to know them; a wet spot, rock ledge, or woodchuck hole is a familiar sight. Year after year, his mind and body are at work in them, plowing, planting, harvesting. A picture of each in his mind, he can recite their soils, drainage, and crop histories. To farm well is to know a ground intimately. Though stretching back to the origins of agriculture, this close contact is now quickly dimming as a mode of perception. Seeing a field is no longer an aching recognition but an antiseptic mapping. What was once in essential ways concrete is today essentially abstract. The most insidious irony of industrial agriculture is that it removes a farmer from his farm; animals, plants, soils, and people are less living things than they are plotted pieces of information.

For example, Iowa farmer Dwayne Gerlach uses an "antique system" to help him plant and harvest his crops: a computer program that makes "precise field maps" (Peterson 32). Using computer-generated maps, Gerlach transforms his fields into abstractions; he knows them as pixels and bits of data as much as he knows them as living places. With more than one farm to run, Gerlach needs a tool that externalizes the wealth of information that he juggles to stay in business. As he points out, his maps save time and "prevent a lot of confusion"; he so relies on them that he keeps them with him "at all times" as he goes about his work. When he gets to his fields, his maps tell him "exactly what the fields look like, the acres in each field, and what seed, insecticide, and herbicide to use"; they show him "exactly what [he's] supposed to be doing." Since buying his computer, he spends "more time in [his] office," but he enjoys the indoors, "thanks to the kind of results [he's] getting." Gerlach uses the same program to design the new machinery that he builds in his farm shop.

Across the state, in northwestern Iowa, Tom Dorr buys state-of-the-art equipment to maintain precision control of his farm. With 3,800 acres and grossing about $2 million, Dorr's farm ranks among the nation's top 4 percent of large commercial farms (Feder, "For Amber" D1). Separating himself even further than Gerlach from his fields, Dorr employs Francis Swain, an information technology expert who sees each field and animal as "simply a source of data." Using U.S. military global positioning systems to collect satellite information, Dorr directs machinery operators "when and where to turn to begin tilling each row of a field"; during planting, sensors monitor tractor speed and "adjust the seed planter." Dorr and Swain dream of creating a 225,000-acre "agricultural factory" that would be "big enough to keep 100-unit trains running to faraway seaports." A central information system would keep the "projected profitability of each field . . . constantly visible to Mr. Dorr, his employees, landowners and the investors he says would be needed to spread the financial risks of such a big enterprise." The men admit that their dream will not soon come true: "Investment bankers have said the project is too small."

## Mining Soil

Industrial agriculture began in the United States as an advertising gimmick. Most Americans first learned about it in the late 1870s and early 1880s through mass-circulation magazine articles that publicized the sudden appearance of large wheat farms in the Dakotas. As prototypes of the New Agriculture, bonanza farms in the Red River Valley of the north introduced "a new style of frontier agriculture, copying the techniques of the rapidly growing American industrial economy" (Drache 68). As "the first factory farms," bonanzas proved that industrial techniques could create a higher producing, more efficient, and more profitable agriculture (91). For a short time the farms were fabulously successful: the original bonanza, the Cass farm, profited over $25,000 in 1878 after only two years in operation (106). But bonanzas treated the land poorly; growing a single crop, they depleted the earth's fertility because they "intensively mined the soil" (219).[1] Though smaller farms displaced them by the early 1900s, their redefinition of farming took firm root in the nation's consciousness, a fact suggested by Frank Norris's *The Octopus*, the first major American novel to understand industrial farming not as a novelty but as a given.

The original Red River bonanza was dreamed up in 1874 as a scheme to demonstrate the viability and profitability of farming Dakota Territory (Benton 406; Drache 42–43). Northern Pacific Railroad promoters created the farm to advertise its large harvests—and profits—in order to lure settlers to Dakota so that the railroad could sell land and recoup financial losses it had suffered in the panic of 1873. Following reports of the farm's success, people flocked to the territory, and the bonanza quickly attracted investors' attention (Drache 72, 74). Soon East Coast capitalists were buying up land and planting wheat, and prominent figures, including President Rutherford B. Hayes, were traveling west to study the new phenomenon (Drache 34, 87 n. 1). Bonanzas were "unique to American agriculture of the 1870's, '80's, and '90's" (Drache 69). Familiar with 160-acre farms, most Americans had no category for a 1,000-acre wheat field—until journalism created one.

As orchestrator of the bonanzas' promotion, railroad land commissioner James B. Power advertised the farms across North America (Murray, "Railroads" 59). In the mid-to-late 1870s, he extolled Dakota farming in articles appearing in large-circulation farm journals such as the *Cultivator and Country Gentleman*. In 1880 *Harper's New Monthly Magazine* and the *Atlantic Monthly*, eastern literary magazines, sent correspondents into the Red River Valley to see what all the fuss was about. By 1897 the farms were so widely known that *Scribner's* targeted them in their series "The Conduct of Great Businesses"; Ray Stannard Baker visited them two years later to describe "The Movement of Wheat" for the muckraking *McClure's* (131). Journalists reporting the bonanza story had to think through something that their contemporaries had never seen: an industrial farm. How they imagined this curiosity points directly to how Gerlach and Dorr imagine their farms today.

Each writer knew that he was witnessing momentous change, a "new movement" in the definition of agriculture, one that was wholly American (Bigelow 33). Poultney Bigelow's "The Bonanza Farms of the West," appearing in the January 1880 number of the *Atlantic Monthly*, heralds "a new development in agriculture, in the great Northwest, [that] has forced itself upon the public attention, that would seem destined to . . . work a revolution in the great economies of the farm" (33). In the March 1880 *Harper's*, C. C. Coffin describes the Red River Valley as the "bread land of the future. . . . The development of wheat culture in Northern Dakota is without a parallel" (529–30). Henry Van Dyke, writing in the May *Harper's*, subscribes the "promise" of the first town he visits to "the vicinity of the gigantic wheat farms, of which all the world has been talking and writing" (805). In an article in the 8 January 1880 *Cultivator and Country Gentleman*, James Power appeals to national pride in describing the bonanzas when he addresses the "people of the United States," who feel a "considerable degree of pride" in feeding the "starving nations of the old countries" (19). To affirm that bonanzas are wholly American, Power, Coffin, and Van Dyke contrast bonanzas with Canadian farm practices; famed Kansas editor William Allen White even boasts that no bonanzas exist on the Canadian side of the Red River Valley, though "upon the American side there is not a barren acre" (532).

Among the first to describe the bonanzas, these journalists define industrial farming in terms taken for granted today: the farms' huge size, heavy capital investment, systematic management, deployment of the latest and best machinery, and potential for accumulating "colossal fortunes" (Bigelow 42). While investigating the New Agriculture, they also note a conflict between old and new conceptions of farming, one that continues today and is best epitomized by the fact that, suddenly, urban capitalists were avid farmers. Most reveled in this, though one writer prophetically points out that the "capitalistic farmer" and the small farmer have "entered into deadly competition," one in which small farmers must lose (Bigelow 43, 42). For this writer, bonanzas signaled the displacement of Jefferson's small-farm agriculture, a displacement other nineteenth-century observers feared would oppress labor, concentrate wealth and power, and hamper the development of rural community life.

While the journalists perpetuated the ancient trope of describing farming as a kind of combat, they did so in the shadow of the Civil War's mass destruction. A bonanza owner does not wrestle, or even battle, with nature himself, as even Virgil claims; he hurls an army against it: "it is just that—the army system applied to agriculture. This general marshals his men, arrays his instruments of war, and with mechanical precision the whole force moves forward to conquer and exact rich tribute from the land" (*Georgics*, Book 1, line 160; Van Dyke 805). The well-known Civil War correspondent C. C. Coffin urges readers to "ride over these fertile acres of Dakota, and behold the working of this latest triumph of American genius" (McKerns 533–34). Coffin's article and its accompanying etchings argue that the war for food is like the war to defeat the rebellion. In other words, propaganda: industrial agriculture appropriates the transcendentally valid image of a national triumph.[2] In one Coffin illustration reapers move in echelon across a boundless field, a squadron of "chariots of peace . . . under the marshalship of this Dakota farmer. A superintendent upon a superb horse, like a brigadier directing his forces, rides along the line" (534). This "brigade of horse artillery" works with the speed and ease characteristic of the New Agriculture: "The grain disappears an instant, then reappears; iron arms clasp it, hold it a moment in their embrace, wind it with wire, then toss it disdainfully at your feet. You hear in the rattling of the wheels the mechanism saying to itself, 'See how easy I can do it!' " (534). Not surprisingly, a general, George W. Cass, owned the first bonanza (Drache 43).

Bonanza battlegrounds were huge: cultivated acreage topped 65,000, a far cry from the Homestead Act's 160-acre farms (Benton 408). Their extent leaves writers grasping for ways to describe them. *Atlantic* writer Bigelow and *Harper's* Van Dyke suggest bonanzas' size by writing pseudotravel narratives, as if each were exploring a New World. Bigelow marvels that one farm has "space enough for three cities like New York" (40). Bonanzas even spawned tall tales that underscored their enormity: " 'A man starts out in the mornin' to plough a furrer, and he ploughs right ahead till night, an' then camps out, an' ploughs back the nex' day' " (Van Dyke 802). Van Dyke's fellow traveler, an illustrator, has difficulty capturing such plowing: "that is the trouble; it's too big. I can't get it on canvas. A man might as well try to paint a dead calm in mid-ocean" (806). Coffin resorts to science fiction: the Red River Valley may be likened to a "great summer wheat field of the future. Its capabilities are so vast, and the insurance of production so certain, that the millions of the Old World may ever think of it as a land that will supply them with bread" (529). William Allen White, an

C. C. Coffin, "Farming in the West—Evening." From "Dakota Wheat Fields," *Harper's New Monthly Magazine* 60 (March 1880): 529.

C. C. Coffin, "Ploughing." From "Dakota Wheat Fields," *Harper's New Monthly Magazine* 60 (March 1880): 530.

C. C. Coffin, "Harrowing." From "Dakota Wheat Fields," *Harper's New Monthly Magazine* 60 (March 1880): 531.

C. C. Coffin, "Sowing the Wheat." From "Dakota Wheat Fields," *Harper's New Monthly Magazine* 60 (March 1880): 532.

C. C. Coffin, "Reaping." From "Dakota Wheat Fields," *Harper's New Monthly Magazine* 60 (March 1880): 533.

C. C. Coffin, "Threshing." From "Dakota Wheat Fields," *Harper's New Monthly Magazine* 60 (March 1880): 534.

anti-Populist and progressive Republican, attributes his readers' inability to see "the bigness of these farms" to the "ghost of the old order"; people whose "preconceived notion of a farm is a little checker-board lying upon a hillside" need help imagining what a bonanza farm is (531, 534). To help them—and astound them—White claims that workmen on a bonanza's individual farms are so scattered that they will not see each other "from season's end to season's end" and that a freight train carrying one farm's crop would stretch two miles in length (534).

Fertile soil made bonanzas fertile fields "for the investment of capital" (Bigelow 33). The farms share with urban industry a Gilded Age entrepreneurial spirit: "Invention, system, capital, brains, are factors for success in most things in these days" (Coffin 531). To handle his bonanza's business, E. W. Chaffee simply created company towns at two Chaffee grain elevators, much as companies often did near eastern coal mines (Drache 145). Seizing the new opportunities, farm machinery manufacturer Cyrus McCormick installed salesmen in the valley; his company made the farms "the major testing ground for the firm's new and improved machines" (Murray, *Valley* 140). By 1880 several "great capitalists," bankers, heads of railroads, and grain dealers had already invested in bonanzas (Bigelow 33; Drache 43, 50). These men were not resident farmers; absentee landlords, they managed their farms from a distance for the greater good of investors, and only incidentally for all who "indirectly profit[ed] by them" (Power 19). Almost overnight, the rich saw farming as worth the risk: no longer could agriculture be only a way of life; in Dakota's fields it became big business for those already making it big in the urban industrial economy.

The new farms operated under a professional, systematic management that appealed to entrepreneurs looking for safe investments. The Cass-Cheney farm, for example, was directed by Oliver Dalrymple, a successful Minnesota wheat grower, whose "system became the model" for other bonanzas (L. F. Crawford 471; Murray, "Railroads" 61). Justifying his articles' statistics, James Power argues that since bonanza owners are "stewards in charge" of a national enterprise, his fellow Americans "expect at least an annual statement of affairs, a sort of yearly accounting, in order that we may assure ourselves that the stewardship is rightly administered" (19). William Allen White ends his 1897 catalog of a bonanza's physical "plant" by mentioning a "set of books, kept as carefully as the books of a bank are kept, and a telephone connecting the farm with a telegraph wire to the world's markets . . . tools of the business—the plant" (538, 540). Bigelow's piece inventories each farm, listing not only the number of acres, but also the number of houses, barns, and sheds needed for a successful operation: "sixty-four harrows; thirty-two seeders of eight feet; six mowers" (39). These would be astonishing numbers to most farmers, but interested investors would certainly appreciate his accuracy—and Bigelow's pleasure in noting that "a Sunday service was held on the Grandin farm, conducted by the book-keeper" (41).

The management structure of a bonanza reflected an industrial division and discipline of labor, another appealing reason for speculators to consider investment, especially for those who feared Populist unrest. Because much of its labor force was unfamiliar with farming, a bonanza's "large armies of men had to be strictly controlled" (Drache 114). Day-to-day operations were directed from farm headquarters; "orders to each division superintendent come from the chief's office the night before for each day's work" (W. A. White 546). Superintendents then passed orders to foremen, who directed workers, who could number a thousand at harvest time (Drache 111). Most workers were "tramps, who vanish when the harvest is over, instead of increasing the permanent population" (Van Dyke 806). In each of Coffin's illustrations a few men supervise the work of several others, a visual representation of industrial farming's undermining of the traditional notion of labor as help, of farmworkers as "social if not economic equals" (Johnstone 147).

As if to ease investors' fears of labor unrest, White praises bonanza farmhands. Noting that of the fifty men employed during the plowing season all but ten are discharged, White claims that the jobless happily "go back to their homes in the pineries or in the cities farther south" (541).[3] These men, "these transient laborers," do the "most important work" (543). And they are not "'hoboes'" but "regular harvesters," "steady, industrious men," who follow the harvests northward, never paying railroad fare, riding into "the bonanza district on the 'blind-baggage' on passenger trains" (543). If these "workmen" have "leisure and a taste for scenery they jolt placidly across the continent homeward-bound in what the lingo of the cult calls 'side-door sleepers'" (543). To assure urban readers that these workers are quite content—at a time when populism's radicalism was a lingering challenge to industrial capitalism—White points out that they "bring home probably a million dollars in wages" (543). Though they may be content, White unwittingly reveals what employers think of them: "the bonanza farmers—at least the better class of them—are as careful of the food set before the men as they are of the fodder that is put before their horses" (545).

These articles place in conflict two visions of farm life: farming as family life, permanence, and community against farming as simply and only moneymaking. Van Dyke juxtaposes bonanzas with a self-sufficient Mennonite community just across the border in Canada whose members work together to reestablish a village, from whose center their "farms radiate" (810). The villagers have found a happy medium between independence and community: "Every man cultivates his own land, and the four-and-twenty families have the advantage of living close together, and making common front against the hardship and loneliness of frontier life" (810). In contrast, each bonanza "absorbs great tracts of land, and keeps out smaller farmers" (806). Whereas the bonanzas import labor and equipment, Van Dyke's Mennonite host is self-sufficient: "All improvements in [his home] the old man intended to make with his own hands at his workbench" (812). Unlike the Mennonites, bonanza owners import their food rather than grow it on their farms (Briggs 32).

Van Dyke and Bigelow agree that bonanzas do not foster communities. Just as many migrant laborers do not find any work after the harvest and must move on, so no family can find on a bonanza a "permanent home by virtue of title in the soil" (Bigelow 41, 43). Jefferson's dream of a nation of freeholders is lost amid the "large development of the tenant system" (Bigelow 44). Bonanzas do not provide space for a concentrated community because they do not wed families to the land: "The idea of home does not pertain to them; they are simply business ventures" (Bigelow 41). Van Dyke contrasts a bonanza superintendent's house, where the sole reported conversation is about the farm's "system of bookkeeping" (806), with a Mennonite's home, where he and his artist friend "seemed to have entered quite into the circle of his domestic life" (812).

Foreshadowing twentieth-century events, bonanza economies of scale shoved smaller farmers out of business. Even farmers with as many as three sections (1,920 acres) under cultivation could not be "counted in the same breath with the more extensive wheat-growers" because the bonanzas "work upon a system of their own" (W. A. White 534). Echoing Bigelow, White understands that the small farmer "cannot economize" during harvest time because he cannot take as full advantage of hired help and machinery as can the bonanza farmer, who can "work the men and the machines to their limits" (545). Bigelow makes a point of contrasting the "shanties of the small farmers" with "the large improvements of the great agricultural adventurers," whose typical farmhouse is "of wood, two stories high, double, painted white, lathed and plastered" (35–36). Shipping huge, steady volumes of wheat, bonanzas also receive " 'special rates' for their transportation" from the railroad, rates "fifty per cent. below the rates charged to the small farmers" (Bigelow 36). To sell at the best prices, the bonanza superintendent "keeps in the closest touch with his agents in the world's great wheat-pits" (W. A. White 547). Competition between small and large farms is most evident in the "distress" of many new settlers; in town, Bigelow hears that "generally the small farmers were hopelessly in debt" (37).

James Power goes to great lengths to refute charges that bonanzas menace small-farm agriculture. According to Power, bonanzas are economically disadvantaged compared to smaller farms: "it costs much more to raise wheat on these great farms than it does the individual who is satisfied to own and work a moderate amount of land, say from 160 to 640 acres" (19). To allay fears that bonanzas were solely for owners' profit taking, Power writes in "the collective first person . . . for it is in a sense 'our' land that produces these magnificent harvests" (19). As if to counter Bigelow's assertion of the lack of community, Power asserts that the valley's population has rapidly increased over ten years. And to encourage more emigration—for the sale of railroad land, remember—Power appropriates the central figure of the agrarian myth to claim that in Dakota, "The husbandman is certain of a rich reward . . . thousands of industrious men, encouraged by the liberal policy of the Northern Pacific Railroad Company, are thronging to the land" (19–20).

While Power may lure settlers with images of agrarian bliss, Coffin goes further to claim that bonanzas herald a new ideal in rural life. For Coffin, pastoral pictures are simply inadequate to represent the realities of late-nineteenth-century agriculture. Because of new inventions, the harvest scene of "reapers bowing down to their work in the golden fields, maidens binding the sheaves," a scene often celebrated by poets and artists, could never be "an American scene" (532). Because of the thresher, "Poets no more will rehearse the music of the flail" (534). The Old World "pictures of the noon rest beneath the branches of an overspreading tree"—family- and community-oriented pictures—are replaced by Coffin's illustrations of entirely male crews "ploughing," "harrowing," "sowing the wheat," "reaping," and "threshing"—crews that are never represented pausing in their work (532, 530–34).

In the New Agriculture women can be removed from farmwork completely, underscoring the new farmer's rise in social class. Celebrating the gender work split that industrial farming hoped to create—men working the fields, women the home—Coffin claims: "True, in the early years of the republic women worked in the harvest field, but invention has dispensed with their labor as followers of the sickle" (532; see Jellison 4). The absence of any women on bonanzas was too obvious to ignore; Bigelow "particularly noticed the conspicuous absence of women and children on the large farms" (41). The only woman pictured in Coffin's article is a well-dressed watcher of a threshing scene, suitably attached to a well-dressed man in a top hat carrying a riding crop, presumably the farm owner. Both look on as over a dozen men thresh wheat (534).

Another component of the new farm is laborsaving machinery. As historians point out, "bonanzas were instrumental in introducing large-scale machinery to American agriculture" (Drache 119; see L. F. Crawford 476). Machinery was not only large but also "uniform . . . so that standard repairs and settings could be applied" (Drache 220). In 1897 Red River Valley bonanzas bought "[t]rain loads" of machinery, "a few thousand dollars less than $3,000,000 worth of machinery annually" (W. A. White 545). The larger farms purchased machinery at a "thirty-three and one third per cent. discount from published prices" (Bigelow 37). And they bought straight from the manufacturer (Drache 119). Many bonanza owners found it cheaper to buy new machinery every year than to fix the old (W. A. White 545). Machinery may have replaced the sickle, but more importantly saving labor also meant working smarter: "The most desirable farmhand is not the fellow who can pound the 'mauling machine' most lustily at the county fair. He is the man with the cunning brain who can get the most work out of a machine without breaking it" (W. A. White 532). But then again, bonanza machinery is so simple to operate a child can handle it (Coffin 532).

Writing in 1897, White expresses a new and powerful argument for farmers to adopt industrial farming: new farmers rely on intelligence rather than muscle. To dispel the preconceptions of "modern men and women [who] pay unconscious tribute" to the old farmer's reliance on "brawn and not upon brain," White claims that the "successful farmer of this generation must be a business man first, and a tiller of the soil afterward . . . the

William Allen White, "Seeding." From "The Business of a Wheat Farm," *Scribner's Magazine* 22, no. 5 (November 1897): 535.

farmer's success in business will quadrate with the kind and quantity of brains he uses, and with the number of fertile acres under his plough" (531–32). According to Ray Stannard Baker, writing a few years later, the "American farmer . . . who plows and reaps and threshes by machinery without so much as touching his product with his hands, is becoming preeminently a man of business" (130). And this farmer is well informed: "The Government has supplied colleges for educating him, and it sends him regular bulletins containing the results of long-continued experiments conducted by the Department of Agriculture. He is a wide reader, sometimes a thinker, and always a politician" (Baker 130). No rustic, Baker's farmer is a sophisticated man of the world.

The bonanza farmer's most important tool was probably a commodity ticker. In contrast to his more self-sufficient and independent contemporaries, bonanza farmers were inextricably tied to a global marketplace —in ways beyond the reach of preindustrial commercial farmers (Baker 130). The commodity ticker expanded the industrial farm to world-size dimensions: "the distance between the fields has been lost. The world's wheat-crop might as well lie in one great field, for the scattered acres are wired together in the markets, and those markets are brought to the farmer's door" (W. A. White 547). Grain farmers a great distance from urban markets benefited most from nineteenth-century advances in the transportation and storage of wheat because "the cost reductions represented a greater percentage of their total costs. . . . Grain farmers near

William Allen White, "Reaping with Right-Hand Binders." From "The Business of a Wheat Farm," *Scribner's Magazine* 22, no. 5 (November 1897): 536.

urban centers were forced to change their specialties" (Schlebecker 134). Bonanza farmers were thus more interested in "the shortage of the wheat crop abroad, and in the steady rise in the price of wheat than he is in the future failure of a soil which for twenty years has shown no 'shadow of turning'" (W. A. White 548). Like Norris's ranchers in *The Octopus*, these men maintained an abstract relationship with nature based on distance. Ignoring the ground they stood on, they worried instead over every price quote that streamed from their tickers.

## At a Distance: Farming with Frank Norris

Industrial farming created a new relationship between farmers and the land, a relationship based on distance and abstraction, rather than on the close bonds and concrete experience suggested in Jefferson's agrarianism or found in preindustrial farming. Implying unlimited financial and physical growth, bonanzas transformed farmers into managers, forcing them from fields into offices. No longer a homesteader, the farmer was not self-sufficient; like his urban counterparts, he ran a complex business knowing he depended on an unpredictable world marketplace. When a farm expands to bonanza size, natural limits cease to be restraints the farmer works within; they become bounds to overcome: deep-well irrigation eliminates deserts, tile draining removes wetlands, chemical fertilizers enhance soil

William Allen White, "Shocking a Sea of Wheat." From "The Business of a Wheat Farm," *Scribner's Magazine* 22, no. 5 (November 1897): 537.

fertility. Physically separated from his farms' expanse, the bonanza owner does not know every foot of his property; he knows and oversees his enterprise secondhand, using employees' reports, experts' studies, and maps: in other words, abstraction replaces experience. Such is the farming in Norris's *The Octopus* (1901). Based on events that were happening even as Bigelow, Coffin, Power, and Van Dyke were describing industrial farming, *The Octopus* is "more a novel about man's relationship to nature than a story of man as a social being" (Pizer 121). Depicting a battle between two major economic forces, a railroad monopoly and a league of bonanza farmers, the novel investigates and puts into prophetic play the characteristics of the New Agriculture.[4]

Norris himself was an urbanite who knew farming only from what he read or observed. A son of wealthy Chicago parents, he studied painting in Europe and English at Berkeley and Harvard (French, *Norris* 21–25). While in New York working for the progressive muckraking journal *McClure's* in early 1899, he conceived the idea for a wheat trilogy, most likely after reading Ray Stannard Baker's account, "The Movement of Wheat" (Waldmeir 57). In a 5 April 1899 letter Norris tells a friend, "I am going to study the whole thing on the ground, and come back here in the winter and make a novel out of it" (Davison 2). He understood the "demands of the New Agriculture" indirectly through extensive research he conducted in California that summer (445).[5] His "chief facts about the lives of the wheat

William Allen White, "Steam Threshers at Work." From "The Business of a Wheat Farm," *Scribner's Magazine* 22, no. 5 (November 1897): 539.

growers" he absorbed during an "extended visit to the San Anita Rancho, near Hollister," which had five thousand acres of wheat under cultivation (Cargill 463; Pizer 124). To experience a wheat harvest, he worked briefly on a harvester-sacking platform (Davison 5; Rothstein 57). After becoming well informed about "every point of view, the social, agricultural, & political" of California bonanzas, Norris returned to New York to write his novel (Davison 2).

Distance defines every corner of *The Octopus*: observers see the ranch lands as "infinite, illimitable" (39); the legal divide separating farmer and railroad is unbridgeable; the poet Presley "dramatize[s] the doctrine of detachment" (Duncan 63); the misogynist Buck Annixter crosses "measureless distances" to marry Hilma Tree (258); the long dead Angéle returns to Vanamee "from the grip of the earth, the embrace of the grave" (276). Images of distance open and close the novel. It begins with a far-off train whistle reminding Presley that he is late getting the mail and that he had "hardly started" on his "long excursion through the neighboring country" (9). It ends with him taking the "larger view," looking back to the San Joaquin from a ship sailing for India with the valley's wheat (458). Apocalyptic news from afar intrudes on social gatherings, redirecting the novel's plot: railroad regrading notices arrive at Annixter's barn dance (194); while at his son Lyman's club, Magnus overhears that the leaguers have lost their court case (228); railroad agents seize the ranches during a ranchers' picnic (356). Donald Pizer notes a similar pattern: "Much of the distinctive effect of *The Octopus* derives from Norris' interweaving of the 'large' and the 'small' . . . a vast epic theme emerges out of the particulars of commonplace individual lives" (161).

Representing ideal new farmers, Magnus Derrick and his neighbor Buck Annixter have no emotional ties to their farms. Both conceive of land only as a physical site for getting rich. Other than financial, Derrick and his son Harran perceive no limits on their uses of nature; their ranch, Los Muertos, "expanded to infinity. There was not so much as a twig to obstruct the view. In one leap the eye reached the fine, delicate line where earth and sky met, miles away" (46). Not born to or trained for farming, Magnus and most of his neighbors run their farms like mines: "They were not attached to the soil. . . . To get all there was out of the land, to squeeze it dry, to exhaust it, seemed their policy" (212). Magnus himself had first grown wealthy in gambling and mining. As for farming, he had simply followed the money (211–12).

A self-made man with "brains to his boots," Annixter is the epitome of the new farmer, a masculine man of affairs who has simply chosen farming to make his fortune (24). Quite wealthy—he inherited a large sum from his land speculator father—his education is the perfect mix of specialized skills needed to transform land and seed into cash: "At college he had specialized in finance, political economy, and scientific agriculture. After his graduation (he stood almost at the very top of his class) he had returned and obtained the degree of civil engineer. Then suddenly he had taken a notion that a practical knowledge of law was indispensable to a modern farmer" (24). To point out that farmers do not soften in universities and that they still measure up to agriculture's physical challenges, the narrator notes that Annixter's "university course had hardened rather than polished him" (24). But with all his education, he "remained one of the people . . . [with] an astonishing degree of intelligence, and possessed of an executive ability little short of positive genius" (24). Like the all-male Red River bonanzas, and unlike Jefferson's family-oriented yeoman, Annixter is a bachelor who has "kept himself free" from "feemales" (26).

## Paper versus Work Values

*The Octopus* is based on the 11 May 1880 Mussel Slough affair in which several farmers were killed attempting to keep a federal marshal from evicting them from railroad land. The Southern Pacific Railroad claimed the land by legal title, while a group of small farmers claimed it by right of toil. The farmers argued that the Southern Pacific was obligated to sell the land to them for $2.50 an acre, the usual federal government price, as soon as the government conveyed land patents to the railroad in exchange for construction of its rail line. Any increase in value beyond $2.50 an acre, farmers insisted, was wholly due to their improvements and ought to be theirs. Farmers further claimed that railroad circulars that enticed them to settle on railroad lands were valid contracts that promised settlers that they would not be charged the value of their improvements. The Southern Pacific Railroad's *California Guide Book* is one such circular; this sentence appears almost verbatim in *The Octopus* (88): "The company invites settlers to go on the lands before patents are issued; and intends, in such cases, to sell to them in preference to any other applicants, and at prices based upon the value of the land without the improvements put upon it by the settlers" (6).

But when the Southern Pacific received its land patents, it moved to sell settlers their farms at the land's *improved* prices, sometimes fifty or a hundred times the original prices named in the circulars. A settler buying land at the improved price, in effect, paid for it twice, in sweat and cash. Mussel Slough settlers interpreted the railroad's regrading as threats: "your money or your life—*your money*, if you have the ready cash; *your life*, if you will but give us a mortgage on your home" (*Struggle* 20). The novel echoes such threats when Harran Derrick tells railroad agent S. Behrman: "My God, why don't you . . . hold us up with a shotgun; yes, 'Stand and deliver; your money or your life'" (56). When the Southern Pacific evicted those who could not pay, the bloody gun battle erupted at Mussel Slough (Robinson 159).

At the center of the novel and the actual events that inspired it is a late-nineteenth-century tension within farming between "paper values" and "work values" (Johnstone 157). Who owned the land: those with an abstract legal title? or those who actually worked and sweated over it? The latter claim was common in nineteenth-century America and had a long history. Andrew Jackson, George Julian, Henry George, and Horace Greeley all maintained that "wild land on the frontier had no value, that its value came from the improvements made upon it and in its vicinity" (Gates 283). In advocating a free land provision for the Homestead Act, Greeley held "that the settler who 'transforms by his labor a patch of rugged forest or bleak prairie into a fruitful, productive farm, pays for his land all that we think he ought to pay' " (Gates 283–84). Greeley imagined a farm worked by a family living in harmony with nature; he could not imagine Dakota's Cass Farm or Norris's Los Muertos.

The title-by-toil philosophy found its political expression most vividly in the late-nineteenth-century Populist movement. Like the Populists, who united under the banner of " 'producerism,' " the belief that "labor creates value," Norris's farmers combine to protect their properties and to obtain a farmer-friendly freight rate to ship wheat on the P & S W Railroad (McMath 51). But Norris's farmers are no Populists. They are not interested in a "cooperative commonwealth" that combines rural and urban protest groups; nor are they keepers of "age-old cooperative labor practices that affirmed the principles of community and equality" (McMath 83, 53). Men after money, they chase principles that earn interest. Though they fight against the railroad's speculative use of land, it is not because they believe, as Populists declared, that land "should not be monopolized for speculative purposes, and alien ownership of land should be prohibited" (Hicks 443). They want land—which they work with tenants and hired men—for their own speculative purposes to secure their "fortunes" (211).

Telling his fortune by the hour, the "most significant object" in Magnus Derrick's office is an urban instrument, a commodity market ticker, whose strip of paper ties Magnus's ranch to wheat pits as far away as Liverpool (44). As the nineteenth century wore on, more and more farmers realized that they were cogs in an intricate system of international markets, accelerating technological change, and capital-intensive business practices. Symbolic of this, Magnus's ticker diminishes his ten thousand acres to "merely the part of an enormous whole, a unit in the vast agglomeration of wheat land the whole world round, feeling the effects of causes thousands of miles distant" (44). The system determined that farmers had to be competitive manufacturers and businessmen who saw adversaries worldwide, not preindustrial commercial farmers who knew their competitors as neighbors. Ceasing to be a complex network of living things, nature became a collection of inputs; land was no longer a home but a factory: "Indeed, the farm of the New Earth is the colossal manufacturing plant of all history" (Harwood 376).

Though a horse-drawn reaper literally separates the preindustrial farmer from his land—the farmer rides instead of walks—he is still in the field, an amalgam of boss and worker. On the new farm, however, management and labor are as widely separated as they are in an urban factory. Annixter's antipathy toward labor erupts when he complains that union men painting his barn are shirking their work (124). He calls Magnus "a fool" for believing he can dismiss all his tenants and work his farm alone (26). Running his farms with a complex hierarchy of workers, Annixter commands like a general in "boots and campaign hat" (94). His "battery lieutenants," his division foremen and superintendents, supervise his farms' fieldwork, while he awaits telephone reports of their progress (94, 97). Communication technologies like the telephone distance the farm owner completely from land and labor by allowing him to command an army of workers from a central location. Directing his farms' division superintendents over the phone, Harran Derrick sits in an office whose "appearance and furnishings were not in the least suggestive of a farm" (43–44). These communication and labor systems, as tentaclelike and controlling as the railroads', make each farm as much an octopus as the P & S W. From Los Muertos it is only a short step to Tom Dorr's Iowa farm.[6]

Not only did it separate the farmer/manager from labor, the New Agriculture separated farmers from their long history of independence. In this New Agriculture, the farmer is a cooperator who often joins with other commodity producers to market products in order to stabilize prices. In the last decade of the nineteenth century, cooperative marketing was on the rise. "There were probably more associations organized between 1890 and 1895 than in all previous years" (Hofstadter 113). Most of these cooperatives failed, but not before several were charged with antitrust violations (Hofstadter 113). Norris's industrial farmers are no different; consciously part of a world market, they too "no longer felt their individuality" and soon realize that cooperation is their only salvation (44). To compete in a world dominated by corporate trusts like the P & S W Railroad, they surrender their independence to organization. To fight a corporate body, the railroad, they create another, the league: "*Organization* . . . that must be our watchword . . . we must stand together, now, *now*" (197).

Whereas Populists organized a political party to lobby for alternatives to the nineteenth century's evolving industrial capitalism, Norris's farmers come together only as "a vague engine, a machine with which to fight" (197; McMath 210). The creation of their League of Defense culminates in Magnus's climactic affixing of his signature to its resolutions (200). But neither he nor Annixter sign up to oppose monopoly per se, as Populists did; Norris's farmers only oppose a particular monopoly, the P & S W Railroad. In a letter written to a friend as he was writing the novel, Norris describes the league's executive committee as "a ring of six or eight men . . . all fairly rich men" (Davison 4). These bonanza owners carry out their bribery scheme alone, consulting "just among themselves" to elect a Railroad Rate Commission that will lower freight rates in the San Joaquin Valley (312; Davison 3). They do not give "a whoop in hell" for the rates charged other farmers (312).

League leader Magnus Derrick even daydreams of forming a farmers' monopoly. Shortly after the league is formed, Magnus meets Cedarquist, the diversified investor and president of the defunct Atlas Company. Cedarquist acknowledges Magnus as a fellow suffering capitalist: "We are well met, indeed, the farmer and the manufacturer, both in the same grist between the two millstones of the lethargy of the public and the aggression of the trust, the two great evils of modern America" (216). Magnus is soon "greatly interested" in the manufacturer's idea that farmers should "do away with the middleman; break up the Chicago wheat pits and elevator rings and mixing houses" and sell their wheat to China themselves (217). When he hears this at an exclusive San Francisco club, Magnus's imagination becomes "all stimulated and vivid"; he envisions farmers "set free of the grip of trust and ring and monopoly acting for themselves, selling their own wheat, organizing into one gigantic trust themselves" (226). He sees himself as its leader, a "pioneer . . . grasping a fortune—a million in a single day" (227). Of course, by this time the upright Magnus has already committed himself to the trusts' worst business tactics by pooling his money with other bonanza owners to buy the railroad rate-fixing commission. Unfortunately, the P & S W outbids him for the one commissioner he counted on most, his eldest son.

While the P & S W Railroad acknowledges ownership only as paper title, the ranchers are mired in inconsistency. Though they assent to ownership as paper title—they do not often assert the "older pioneer and agrarian notions" of "actual possession and use"—the ranchers cloak themselves in title by toil principles when it suits their purpose (Johnstone 157). Annixter, for example, adheres to paper values; hearing a rumor that the railroad will finally affix a value to the disputed ranch lands, he argues that the irrigation ditch is a capital improvement: "Why, Magnus and I have put about five thousand dollars between us into that irrigating ditch already. I guess we are not improving the land just to make it valuable for the railroad people" (74). But later, when the railroad refuses his check to purchase the disputed land, Annixter falls back on work values: "Who made it worth twenty [an acre]? . . . I've improved it up to that figure" (141). During the gunfight, the irrigation ditch—representing to ranchers their title-by-toil claim—is a "natural trench" from which they hope to halt railroad attempts to appropriate lands each asserts to have improved himself (359). When Annixter leaves the safety of the ditch to confront those holding paper title to his land, Osterman shouts at him, "keep in the ditch. They can't drive us out if we keep here" (366). Unfortunately, the ranchers die after they leave the security of their most significant grounds (366–67).

Only a commodified nature emerges unscathed from the novel's havoc: "*the WHEAT remained*. Untouched, unassailable, undefiled" (458). Reducing farming to manufacturing—as Magnus and Annixter do—manifests a false hope that the living natural world can be well managed—at a profit—and that humans can precisely draw the necessary boundaries. But nature has a nettlesome way of refusing reduction. A recent—and telling—example from today's industrial farming is mad-cow disease, bovine spongiform encephalopathy, a new and fatal brain disease of cattle that is likely transferable to humans through ingested beef (Hill et al. 448; Hillerton 3042). In the 1980s British cows were infected by eating feed containing ground-up sheep and other animals that had contracted the disease (L. M. Crawford 12). The mind behind the feed, of course, erected barriers between sheep and cattle and cattle and humans that it thought impermeable. This assumption killed several people and thousands of animals and rocked the British economy. Only an agriculture that defines animals as data could imagine feeding meat to a herbivore in the first place.

## Managing Nature

Norris's ranchers live and die by their improvements, by their efforts to order and control nature. The climactic, bloody battle takes place at the irrigation ditch, but more, the ditch is an improvement underwriting others and it is the one that has the greatest potential to raise land prices even further—and land price is the root of contention in the novel. According to the map that opens *The Octopus*, the ditch is a significant undertaking, nearly five miles long. Its very presence signals that these farmers are not independent yeomen; built by two new farmers, it serves to minimize risks posed by drought. A labor- and capital-intensive project, the ditch was probably surveyed by civil engineer Annixter and is constructed by an army of hired hands. Such large-scale irrigation denies nature's rhythms and transforms farming into a "factory operation" (Worster 126).

Like any farmers, Norris's ranchers fashion an artificial landscape. But theirs is a vast, rationally and systematically constructed one, its wildness reduced to watercresses under a railroad bridge or jackrabbits rounded up for slaughter (229, 353). As pictured in the novel's map, Norris's bonanza farmers inhabit a landscape defined by precisely planted wheat fields, barbed-wire fences, farm divisions, and railroad tracks. The land itself is later described as an abstraction, a map, unrolling before Presley like "a huge scroll" (39). Such landscapes are not preindustrial or pastoral homesites; they are simply fertile locations for gigantic businesses. To assert that the novel is "essentially pastoral in its movements and harmonies" because of its landscapes is to miss Norris's point (Graham 110). His genius is imagining at the outset the calamitous results of remaking farming as urban industry. His San Joaquin bonanzas are factories in the field.

Preindustrial farmers sought to reduce nature as a risk factor in production; they drained fields, planted improved crop varieties, and adapted machinery to landforms. In contrast, industrial farmers hope to remove natural processes from production completely, creating hydroponics, hybrid corn, seedless oranges, and tomatoes genetically engineered to fit harvesting machines. The most extreme modern example is cloning. Erasing nature became a farmer's dream when natural processes became line-item expenses on the farm's balance sheet. Whereas before hidden costs associated with natural cycles were usually reckoned in the aggregate, accounting systems borrowed from urban industry inform new farmers at a glance where potential profits go. In *The Modern Farmer in His Business Relations* (1899), Edward Adams notes that "evolution has at last developed a race which, having overcome all other beings, shows signs of trying conclusions with nature herself" (127). The biggest bonanza would be the one where the wheat produced itself.

Whereas preindustrial agriculture adapts to natural forces at a local level, industrial agriculture tries to overcome nature on a limitless scale. It is not surprising, therefore, that land alteration causes the railroad/rancher trouble in both the novel and the incident on which it is based. Because central California is semiarid, San Joaquin farms need a reliable source of water. Mussel Slough farmers and railroaders fought over improvements made possible by irrigation, and the same happens in the novel. As it remains today, water was power in nineteenth-century central California. In the Tulare Lake Basin, in "the early 1870's, irrigation was recognized as a method by which farmers could gain more control over their domain and, at the same time, vastly increase the value of their property" (Preston 136).[7] In news accounts of the Mussel Slough shootings, documents that Norris must have read in his research, irrigation ditches are named as the creators of property values (Pizer 123, 194 n. 23).[8]

Underscoring the importance of irrigation in the novel are constant references to extreme drought conditions in its first chapter. For instance, as the novel opens the land is so dry that "the layer of dust had deepened and thickened to such an extent that more than once Presley was obliged to dismount and trudge along on foot, pushing his bicycle in front of him" (9). Broderson Creek is "a mere rivulet running down from the spring . . . Mission Creek on Derrick's ranch was nothing better than a dusty cutting in the ground" (37). Harran Derrick reflects that the drought had brought ranch tenants like Hooven to "the stage of desperation . . . a third year like the last, with the price steadily sagging, meant nothing else but ruin" for the entire ranch (46). Hooven hopes to remain on Los Muertos to care for the ranch's watering needs: "Who, den, will der ditch ge-tend? . . . [and] der pipeline bei der Mission Greek, und der waater hole for dose cettles" (11). A working irrigation system continually supplied with water is the rancher's only hope to overcome rainfall's unpredictable role in production.

To erase risks associated with rainfall, the ditch will carry water from Annixter's artesian well, rather than unreliable runoff from mountain snows or from the diversion of nearby streams. The well locates the ranch, just as the live oak marks Hooven's home; when Presley first gets to Quien Sabe he notices "Annixter's ranch house and barns, topped by the skeletonlike tower of the artesian well that was to feed the irrigating ditch" (23). An artesian well involves deep drilling to tap an underground stream so that water rises to the surface by its own hydrostatic pressure, thus becoming a "natural" flow. The flow can be regulated mechanically, ensuring that the land gets just the right amount of water to nourish a crop properly. Artesian wells rarely run dry, though the aquifers they often tap are now threatening to (Worster 127).

An urban irrigation device represents Los Muertos's prosperity. When readers first meet Harran Derrick, he is "in the act of setting out the automatic sprinkler" to keep the Derricks' lawn alive (12). An automatic system frees Harran from close contact with the lawn's care, leaving him time to contemplate and appreciate it at his ease. Later, identifying himself with an urban command of nature, he expresses pride in this lawn: it "was as green, as fresh, and as well-groomed as any in a garden in the city" (46). But nature reasserts itself when the Derricks lose their bonanza: "grass on the lawn was half dead and over a foot high; the beginnings of weeds showed here and there in the driveway" (436). Unsupplied with managed water, their urban lawn quickly gives way to California's natural semiarid conditions, a lesson that resonates today.

## Nature on Paper

Because private property requires a method and record of division and assessment, we use maps to represent how we parcel, organize, and value nature.[9] For example, the U.S. survey grid system, centrally controlled from Washington, D.C., "forced a geometry of squares and rectangles upon the naturally curving field shapes based on terrain. . . . The square and rectangular survey had no final boundary line" (Opie 3). Designed by Thomas Jefferson, the grid system was supposed to move public lands into private hands in an orderly manner. Theoretically, settlers first saw their properties on a land office map: land was an abstraction before it was a particular place. Thus many homesteaders knew their land as alienable long before they developed an emotional attachment to it. But maps are especially important for bonanza farmers for reasons aside from their representation of property rights. Because bonanza owners cannot know thousands of acres intimately, they rely on maps to conceptualize their farms and to plan systematically what to do with them. Paper maps allow a farmer the illusion of reducing a bonanza to his vest pocket.[10]

Like any magazine article or novel, a map is embedded in a rhetorical situation; each is the intersection of maker, reader, subject, and purpose. Though masquerading as a disinterested view of the world, a map in fact asserts the mapmaker's perception. Maps never simply copy reality—they shape it. Abstractions, they reduce living things to data. The novel's five maps illustrate the tension between the concrete and the abstract that runs throughout the text; taken together, each marks an increasingly abstract representation of nature that draws the reader further and further from the ranch lands. Norris suggests their importance in a September 1900 letter, written when he was nearly finished with the manuscript: "In the front matter I am going—maybe—to insert a list of dramatis personae and—This *surely*—a map of the locality" (Davison 6). The first map follows the list of characters; it is a "map of the country described in 'The Octopus'" (8).

This map reduces nature to trees, creeks, springs, high and low ground; a property map, it defines the country with criss-crossing straight lines representing ranch divisions, telephone lines, rail lines, and roads. The novel's first line orients readers to roads that follow grid survey lines: "the County Road . . . ran south from Bonneville and . . . divided the Broderson ranch from that of Los Muertos" (9). What follows describes the poet Presley mapping the landscape from a bicycle, which limits his travel to these roads. In effect, readers are introduced to the bonanzas as urban passersby; during the bicycle craze in the late 1890s, the two-wheeler was "mainly associated with urban and newly suburban ways of life" (Ellis 18). With a glance at the visual map, readers see a commodified countryside whose story Presley's verbal map unfolds. The story and the map offer readers the new farm fantasy of fully knowing "measureless" agricultural landscapes (39).[11]

Used solely in wheat planting, the second map is Magnus Derrick's window on his land; he is saved the trouble of walking his fields to know them. The map also erases the possibility of other crops, animals, human habitations, and people. Covering a wall in Magnus's office, this "great map" details Los Muertos's topography (43). A campaign map on which he plans his assault upon the land, the map depicts "every watercourse, depression, and elevation, together with indications of the varying depths of the clays and loams in the soil, accurately plotted" (43). The eye of "the nerve center of the entire ten thousand acres of Los Muertos," the map reassures Magnus in his illusion of control over natural processes (43).

Hanging in the P & S W land office is the railroad's assertion of its property rights. This map denies that a struggle exists between farmers and the railroad over land title; it neatly separates railroad and rancher lands by delineating "the railroad holdings in the country about Bonneville and Guadalajara, the alternate sections belonging to the corporation accurately plotted" (139). As Annixter sits in the office offering to buy land he thinks is his for $2.50 an acre, he cannot fail to see over the desk the "vast map" on which his ranch is platted as railroad land (139). The map denies him property he believes is his; it mocks him as he tells the railroad land officer that he wants to "feel that every lump of dirt inside [his] fence is [his] personal property" (141). But in the next breath he tacitly admits the map's validity when he says that his "ranch house—stands on railroad ground" (141). In effect, Annixter is helpless before the map's power to shape reality because he assents to the map's assumption that land is solely a commodity. He makes this belief very clear: he wants outright ownership of all the land he farms just in case he wants to sell it to "take advantage of [its] rise in value" (141).

The fourth map, which Lyman Derrick receives at the opening of Book II, represents the P & S W system as a living organism feeding on California. Signaling to Lyman his newfound power as railroad commissioner, this "official railway map of the State of California" is also "accurately plotted" and depicts in various colors all railroad lines in California (204). P & S W rail lines, drawn in red, "gridironed" the entire state, crowding out all other colors/railroads (204). The map envisions the octopus for which the novel is named: the P & S W is a "plexus of red, a veritable system of blood circulation" that sucks the "lifeblood" from the state; with a "hundred tentacles," it is a "monster . . . gorged to bursting; an excrescence, a gigantic parasite" (205). "Consumed by inordinate ambition" to be governor of California, Lyman sees in the map not only the state he wants but also his only line to get it (212). Having "outlined" his career twenty years ahead to govern California, he must exchange family blood for company blood; a liar, Lyman works with "the unshakable tenacity of the coral insect" to realize his goal—to achieve his dream he sells out his father, Governor Magnus (213).

On Lyman's map ranches disappear into its white space, just as they disappear completely at the end of the novel, when they are imagined from out at sea, where one can only know one's position from the stars. The *Swanhilda*'s master plats for Presley the novel's most abstract map: "The land yonder, if you can make it out, is Point Gordo, and if you were to draw a line from our position now through that point and carry it on about a hundred miles further, it would just about cross Tulare County not very far from where you used to live" (456). In the end, the ranches are lost to view, locatable only by navigation coordinates and a memory inspired by looking at "the faint line of mountains that showed vague and bluish above the waste of tumbling water" (456). Abstraction wins. Though the farms' concrete commodity remains—the *Swanhilda*'s hold is filled with a bonanza crop of wheat—the poet Presley can imagine it only as a "mighty world force" (458).

## Art and the New Farm

Working in perfect harmony with the New Agriculture, the novel's artists hide its realities with a broad bucolic brush. Confining a pastoral, preindustrial agriculture within its frame, Hartrath's landscape painting is an empty, sentimental piece displayed only among the urban wealthy. "A Study of the Contra Costa Foothills," a landscape sketch, depicts "reddish cows, knee-deep in a patch of yellow poppies . . . [and] a girl in a pink dress and white sunbonnet" (220). Raffled off at an exclusive male club in San Francisco, the painting directs wealthy urban women's attention away from the new farm world; the painting's agricultural subject feeds their imaginations with a pleasant rural world that does not exist. One of Mrs. Cedarquist's "fakers" (221), Hartrath bounds his pastoral scene in "a frame of natural redwood, the bark still adhering" (220). By trying to mask the artificiality of his representation of farming, Hartrath merely succeeds in pointing it out, especially for those familiar with the New Agriculture. His limited, static landscape contrasts starkly with the expansiveness of the novel's bonanzas. Underscoring its emptiness, the painting's female admirers judge it in the abstract, dilettantish language of art books and painting classes: "They spoke of atmospheric effects, of middle distance, of chiaroscuro, of foreshortening, of the decomposition of light, of the subordination of individuality to fidelity of interpretation" (220). Another dilettante, Mrs. Cedarquist, who heads a charity to relieve a famine in India, wins the painting at the moment when Magnus Derrick finds out that the League of Defense has lost its court case against the railroad. While the urban audience's attention is on Mrs. Cedarquist shouting "I've won I've won," the stunned and disheartened farmer leaves the club unnoticed (228).

Hartrath may be arrogant and opportunistic, but Presley is self-deluded and contradictory. Searching for "romance," the "noble poetry of the ranch," this urban bard finds among the bonanzas only "realism, grim, unlovely, unyielding"; his wish "to see everything through a rose-colored mist" separates him from the particular problems of particular people (15). Like the New Agriculture's view of things, his artistic vision is merely farsighted, reducing everything near at hand to "only foreground, a mere array of accessories—a mass of irrelevant details" (39). Obsessed with writing an epic about the "huge romantic West" that would "embrace . . . a whole epoch, a complete era, the voice of an entire people," he cannot see individual people as people: "These uncouth brutes of farmhands and petty ranchers. . . . Never could he feel in sympathy with them, nor with their lives, their ways" (15, 13, 10). He has trouble writing his poem because he can only imagine it finished; like Magnus, who sees only "the grand coup, the huge results," he cannot be troubled with the details of its production (227). A literary bonanza, his epic will encompass all "the traits and types of every community from the Dakotas to the Mexicos . . . welded and riven together in one single, mighty song" (14). And though he wants to write for the people, the everyday and the commonplace "irritated him and wearied him" (15). When he finally does compose a poem, "The Toilers," he writes at the people rather than for them.

For all his concern about reaching "the people," Presley believes that publishing his work in the "daily press" would be "throwing [the] poem away"; yearning for fame, he wants readers of "monthly periodicals, the rich," to see his work (265). When scolded by Vanamee for his selfishness, Presley abruptly decides to publish "The Toilers" in a daily paper, although he does so more to prove to himself and to his friend that he is sincere than to be a "helper of the helpless" (265–66). When he gets another chance to reach "the people," he fails miserably; his impassioned speech at the Bonneville rally comes off as too "literary," and as he leaves he knows that he remains an "outsider to their minds" (389). Not coincidentally, the next time we see him he is visiting Cedarquist in his San Francisco office (395). Back in the city, content to play the esthete, Presley feels more at home at his downtown club or at the Gerards' Nob Hill mansion than in the streets among people like the starving Mrs. Hooven.

If Presley's poem alludes to Edwin Markham's poem "The Man with a Hoe," as critics claim, then Presley remembers a preindustrial agriculture rather than face the new farming that kills his friends (Cargill 466). Markham's poem, published in January 1899, found its inspiration in Jean-François Millet's painting *The Man with a Hoe*, which depicts a French peasant stooped over a hoe in a field he works alone—hardly a portrait of industrial farming. Markham's poem, a challenge to "lords and rulers in all lands," is as easily an indictment of bonanza owners as it is of urban industrialists. Presley's momentary socialist sympathies rightly lead him to damn inhumanity, but the inhumanity he criticizes he remakes: the cruelty is originally not between classes but within one. Diverting attention from bonanza money and power, his poem appeals to the rich, who politely compliment the poet while they all dine sumptuously, easing their consciences with gossip about famine relief—even as Mrs. Hooven stumbles through the streets (426). When Shelgrim, ruler of the railroad, concludes that the painting is better than the poem—and that the poet should have been silent in the first place—Presley quickly agrees, noting that Shelgrim understands the "ethics of a masterpiece of painting" (404). The railroad boss likes the painting because, like Hartrath's, it keeps before a nostalgic public a rural landscape that is nothing like the industrialized landscapes that he has just snatched up. As Cedarquist points out, Shelgrim actively supports art festivals like the Million-Dollar Fair because "it amuses the people, distracts their attention from the doings of his railroad" (224).

Presley's aesthetic distance remains unchanged after all he experiences in horrific detail. Just as in the beginning he fancies himself "[a]s from a point high above the world, [where] he seemed to dominate a universe, a whole order of things" (39), in the end his perspective has hardly changed. Looking back to the San Joaquin from the *Swanhilda*'s deck, he dismisses the novel's violence and greed by lamely declaring that men are "motes in the sunshine" (457). Sobered but not changed by the chaos around him, he simply leaves the country, finding refuge in an ordered universe commanded by a "larger view" that reasserts the irrelevance of detail and excuses maliciousness: "Greed, cruelty, selfishness, and inhumanity are short-lived; the individual suffers but the race goes on . . . all things surely, inevitably, resistlessly work together for good" (458). The San Francisco poet absorbs this reassuring philosophy from the wandering Vanamee, the novel's other poet, a man as unconcerned with detail as Magnus or Presley himself.

Cloaked in the language of pastoral romance, Vanamee's "fancy" represents the most extreme form of the New Agriculture's zealous drive to usurp natural processes (108). In an "isolated garden of dreams, savoring of the past," Vanamee really believes that he can "concentrate [his] power of thought" to return his lover, Angéle, from the dead (109). Joining a bonanza gangplow after being fired from his shepherd's job, this "poet by instinct" reaches beyond the grave to reproduce their blissful moments together; the former shepherd trains his extrasensory powers until his angel is "realized in the wheat" (32, 449). But his new faith in a "large enough" view of things, that "life simply is," a view that looks out upon all "from the vast height of humanity," brings him only his lover's daughter (447). In insisting that the elder Angéle has returned to him, Vanamee refuses to see the woman before him. Satisfied with what amounts to a figment of his imagination, the "half-inspired shepherd-prophet of Hebraic legends" will not discuss with Presley why he is so happy (446): "Do not ask me any further. To put this story, this idyll, into words, would, for me, be a profanation. This must suffice you. Angéle has returned to me, and I am happy. *Adios*" (447). He believes his "delusion . . . the miracle," even if it is "dementia" (108). To speak of it would restrain his illimitable vision, forcing him to account for its concrete details, thus destroying it.[12]

## Masculine Farm/Feminine Farm

Norris describes the divide between old and new conceptions of agriculture in terms of the "masculine-feminine ethic" that many have noted in his fiction (Pizer 87). For Norris, masculinity meant "action and force rather than thought or sophistication"; men were doers in a world of constant struggle (Pizer 94). The proper role of a woman was "to exercise a tranquilizing and ennobling influence upon her husband" (French, *Norris* 87). Women "were to be protected; they were expected to live unselfishly for their men" (Dillingham 88). Norris's ideal woman was a " 'man's woman,' " someone who embodied a "strength and seriousness of purpose," a woman who helped a man "to strengthen his moral courage and his will to achieve" (Pizer 88, 87). His more " 'feminine' " women, however, are "cringing little creatures who are afraid of life" (Dillingham 88). The central women in *The Octopus*, Annie Derrick and Hilma Tree, do represent Norris's divided view of women, but both are associated with pastoral conceptions of farming that must be crushed in a new farmer era. In the logic of the new farm, as C. C. Coffin and Poultney Bigelow point out, women have no place.

Norris portrays the essence of the New Agriculture itself as masculine. For example, while plowing, Vanamee imagines thousands of plows "upstirr[ing]" the San Joaquin Valley, "tens of thousands of shears clutched deep into the warm, moist soil" (96). The fall plowing was "vigorous, male, powerful, for which the earth seemed panting"; only something the size of the New Agriculture's work system could hope to woo "the Titan" earth (96). In Norris's San Joaquin, the new farming and the earth are "world forces, the elemental male and female, locked in a colossal embrace" (96). New farming takes men, potent men like Magnus Derrick. Embodying "the spirit of the West, unwilling to occupy itself with details, refusing to wait, to be patient, to achieve by legitimate plodding," Magnus saw husbanding resources as "niggardly, petty, Hebraic" (211–12). Losing his bonanza unmans him; his "old-time erectness was broken and bent. It was as though the muscles that once had held the back rigid, the chin high, had softened and stretched" (437).

The novel's most "feminine" woman, Annie Derrick, is the perfect new farmwife: with little work to do, she confines herself to a dollhouse, restricting her mental life to memories of a preindustrial agriculture and to an effete literature that Norris despised. A shy, frightened woman, she is not the "reservoir of moral strength" for her husband that Norris believed the ideal woman should be (Pizer 112). Annie Payne Derrick is no "man's woman"; she has the "eyes of a young doe" (48), eyes that "easily assumed a look of inquiry and innocence, such as one might expect to see in a young girl" (47). Filled "with an undefinable terror" of her husband's bonanza, she cannot accept the "new order of things"; for her "there was something inordinate about it all, something almost unnatural" (48).

To get her bearings, she nostalgically recalls her childhood's Ohio homestead—through a misty lens that Horace Greeley would have approved. Her memory of her parents' farm has all the markings of the ideal farm of the agrarian myth: "cozy, comfortable, homelike; where the farmers loved the land, caressing it, coaxing it, nourishing it as though it were a thing almost conscious; where the seed was sowed by hand, and a single two-horse plow was sufficient for the entire farm; where the scythe sufficed to cut the harvest, and the grain was thrashed with flails" (48). Here, Annie Derrick daydreams about living close to nature on a self-sufficient farm, her family growing a diversity of crops using simple farm tools operable by one person. But on Los Muertos, her past has no place. Though the well-defined farm of her imagination is immediately apprehensible to her eye, the illimitable expanse of Los Muertos is unknowable; "never for one moment since the time her glance first lost itself in the unbroken immensity of the ranches had she known a moment's content" (48). To survive, she "retired within herself," living within the manageable limits of "a little summerhouse" built in "the lower branches of a live oak near the steps" outside Derrick's big ranch house (48, 46).

Baffled by the world besieging her, Annie Derrick also retreats to a literary education that has not oriented her to the realities of bonanza life. Her inability to locate herself within the vast acres of wheat forces her to find order in "poems, essays, the ideas of the seminary at Marysville"; amid the "neatly phrased rondeaux and sestinas and chansonettes of the little magazines," she nourishes in safety her anachronistic assumptions about farm life (49). Her literary tastes contrast with the more vigorous ones of Annixter, who reads *David Copperfield* over and over; she is revolted by the "nobility and savagery . . . heroism and obscenity" of Presley's proposed Song of the West and Homer's epics (49). Rooted in a static, pastoral vision of life, she eventually returns with her broken husband to her former life as an English instructor (437). As a teacher of imaginative literature, she will pass to the next generation her pastoral vision of farm life, rather than the violent reality of her husband's bonanza.

The New Agriculture is not content to ignore Annie Derrick's nostalgia; this wide and busy farm world shatters her introspective existence. Magnus's bonanza simply overwhelms her; she is "frightened" by Los Muertos, a farm "bounded only by the horizons . . . ruled with iron and steam, bullied into a yield of three hundred and fifty thousand bushels" (48). "Stunned" by the ranch's monoculture, she cannot cope with the "direct brutality of ten thousand acres of wheat, nothing but wheat as far as the eye could see" (48). Her constant fright marginalizes her; at Annixter's barn dance, for example, she "settled herself in her place in the corner of the hall, in the rear rank of chairs . . . glad to be out of the way, to attract no attention, willing to obliterate herself" (172). Though she tries to restrain her husband from committing himself to the ranchers' bribery scheme, the crowd casts her aside like a "feather caught in the whirlwind" (199). After Magnus joins the league to fight the railroad, Annie is "swept back, pushed to one side" by the crush of farmers rushing to congratulate her husband (200). She does not have the strength to keep him from abandoning his code of ethics—which leads to his downfall and to her "settled grief" and "despair" (437).

In contrast to Annie Derrick, dairymaid Hilma Tree is a "man's woman." Described as possessing a "healthy, vigorous animal life" and a "rigorous simplicity," Hilma transforms Annixter from selfishness to generosity (64–65). While the bachelor Annixter may work with nature abstractly and at a distance, Hilma possesses the "natural, intuitive refinement of the woman not as yet defiled and crushed out by the sordid, strenuous life struggle of overpopulated districts. It was the original, intended, and natural delicacy of an elemental existence, close to nature, close to life, close to the great, kindly earth" (65–66). Her "elemental existence," her pastoral life in the dairy, makes her the perfect woman to act as an ameliorating force in Annixter's life. The tragedy of his death is that their marriage of old farm and new farm values does not last long enough to effect change throughout the community. As if to underscore this, she suffers a miscarriage the night of his death (378).

Annixter's reorientation hinges on the point when he thinks of others, not himself; his new attitude contrasts sharply with Magnus's characteristic new farmer attitude toward others: "After us the deluge" (211). The moment occurs after he realizes that he needs to "think less and feel more," when he turns from abstract to experiential knowing: "Marriage was a formless, far-distant abstraction, Hilma a tangible, imminent fact" (258). Marriage and Hilma soon fuse; like "the harmony of beautiful chords of music, the two ideas melted into one" (259). Only after the union of the abstract idea and the concrete woman can Annixter live an unselfish life, feeling for others, joying in the "boundless happiness in this love he gave and received" (350). Whereas before he was a new farmer "whose only reflection upon his interior economy was a morbid concern in the vagaries of his stomach," loving and working with Hilma "changed [his] life all around" (26, 284). A new man, he tells her: "I'd been absolutely and completely selfish up to the moment I realized I really loved you" (284). Knowing love for the first time in his life, he acts with a generosity and restraint that cannot be permitted to last in a new farm world; tragically, giving up his new farmer attitude kills him. Acting with restraint at the irrigation ditch, he even surprises the other ranchers by declaring that the confrontation should be "settled peaceably" (360). He faces the railroad's hired guns not for himself and his financial interests but for Hilma and his unborn child. During the ensuing battle he is "instantly killed" defending the home that Hilma made possible, a place that before had been "an office where business was transacted" (368, 117).

## Larger Demands

Industrial farming is the sole winner in the novel's artificial landscapes; ranchers are killed or dispossessed, displaced tenants starve in San Francisco, ruined rancher Dyke is hunted and killed. After the gunfight and so many ranchers' deaths, the scope of wheat production only enlarges: Broderson's buildings, for instance, "were being remodeled, at length, to suit the larger demands of the New Agriculture" (445). Eliminating unnecessary farmers, the gun battle looks forward to later cries that there were simply too many farmers. Even triumphant railroad lackey S. Behrman, who fancies himself "the master of the wheat," suffocates in Los

Muertos's bonanza crop, the wheat "a resistless, huge force, eager, vivid, impatient for the sea" (433, 451). Norris here underscores the New Agriculture's victory by personifying the WHEAT as an abstract force, one existing beyond nature, a "mighty world force, that nourisher of nations, wrapped in Nirvanic calm" (458). By novel's end, industrial agriculture has reached its logical conclusion: "No human agency seemed to be back of the movement of the wheat" (451). From field to harvester to railcar to ship, the wheat "moved onward in its appointed grooves" in a vertically integrated system anticipating today's agribusinesses (458).

Pitting two systems against each other—the railroad and industrial agriculture—*The Octopus* shows that complex divisions of labor, communications, and economies of scale invite disaster. Growing ever bigger, such systems get unwieldy, so much so that local mischance can quickly mushroom to catastrophic failure. For example, in 1975 a "fire-retarding chemical" mixed with livestock feed led to the destruction of over a million chickens, nearly thirty thousand head of cattle, and thousands of hogs and sheep; the cause " 'may have been as simple as pulling the wrong lever' " (Berry, *Unsettling* 222–23). In December 1984 a Union Carbide plant manufacturing pesticides in Bhopal, India, leaked forty tons of noxious gas into the city, killing 3,500 people, injuring over 150,000, and sending 400,000 fleeing in panic (Shrivastava 254). This brutal industrial accident was caused by, among other things, a "[f]aulty 'pipe washing' procedure" and the "simultaneous failure of four safety devices" (Shrivastava 254–55). More recently, in July 2000 over 12,000 Japanese were sickened from bad milk caused by "a bacteria-laced valve on a milk storage tank that had not been cleaned as frequently as required" (Takada; see Sims and Yamaguchi). As if mimicking in its form the complicated agricultural system it portrays, *The Octopus* piles up so many themes and structures—and events, characters, and symbols—that it too becomes unwieldy—to the brink of incoherence (Duncan 56–57; Sarver 76).

The remoteness of Norris's farmers from nature foretells farming's increasingly abstract relationship with the natural world since the beginning of the twentieth century: farmers like Dwayne Gerlach and Tom Dorr not only ride in air-conditioned tractors but they also use computers and satellites to plant crops. The P & S W Railroad's easy absorption of the novel's bonanzas prefigures the swift, late-twentieth-century vertical integration of farming; today, only a handful of companies dominate large segments of agriculture: Monsanto and Conagra are examples. Another, chicken giant

Tyson Foods, crows to its investors that it "oversee[s] every aspect of production and control of [its] products' quality and cost from the egg to the dinner table" (Tyson). These conglomerates dominate today because the pressures of nineteenth-century industrial capitalism forced agriculture to become ever more mechanized and capital-intensive, to become ever more productive and less labor-intensive; around the turn of the twentieth century, farming metamorphosed into an industry that valued efficiency, unlimited growth, and unrestrained mining of nature. As the conflict escalated over who would benefit most from this exploitation, farming became progressively harmful to the environment, less and less decentralized, and more and more farmer killing.

Norris was eerily prophetic. As his industrialist Cedarquist asserts with deadly accuracy, the farm problem of the twentieth century was markets—unfortunately, those markets ripped most farmers from their farms (216). Conceiving *The Octopus* as the first in a trilogy, Norris hoped to trace a crop of wheat from its "production" in California to its "distribution" through Chicago to its "consumption" in Europe as famine relief (vi). His second novel, *The Pit*, published posthumously in 1903, tells the story of a speculator ruined in the Chicago wheat pits. Does Norris's third novel nod to the twenty-first century: will this century's farm problem be consumption—in an increasingly hungry world, a world without farmers, a world whose fossil fuels are fast disappearing? If the recent famines in Ethiopia, Somalia, and North Korea are any indication, the answer is yes. Though Norris died before his third novel could be written, *The Wolf*'s title is too suggestive to be ignored.

The twentieth century erased the American farmer. With the individual farmer gone, important questions must be asked. Will the United States quit producing food and import its sources of life? Some think this an inevitable—and wonderful—idea (Blank 3). Will corporate food producers keep chemical fertilizers from polluting streams and rivers, stop insecticides from killing off more species of plants and animals, halt topsoil from disappearing at an alarming rate? Their track record suggests not. We must reorder our thinking: our only alternative to sure destruction is living closer to nature, not farther from it. But any rethinking must be accompanied by a reworking. We need to roll up our sleeves, return to our fields, and imagine nature as our measure rather than as simply ours (Jackson 73).

**Vo Van Tam, with his wife, is a farmer in Song Phu village, where rice yields have risen 500 percent in recent years.—*Washington Post*, 15 May 1994**

CHAPTER TWO

# Challenging the Agrarian Myth

## Women's Visibility in the New Agriculture

On Saturday, 4 June 1994, under the heading "Invisible but for the Picture," the *Washington Post* printed letters from two readers indignant over a front-page *Post* photograph. Accompanying an article about Vietnam's growing economy, the picture depicted a man and a woman standing beside a rice field. Its caption identifies the man as a farmer and the woman as "his wife." The letter writers vigorously contend that the photo and article erase the woman's farming role by referring only to the man as the farmer. As the first letter writer puts it, "one of these people has benefited from new land distribution policies; one of these people farms about an acre and a half of rich land, one of these people has contributed to soaring rice yields in the region" (Cochran). The other has not. But both work the land and both reap the rewards of their labor (Johnson). So why, the letter writers ask, is she invisible but for the picture?

Decades earlier feminist writer Charlotte Perkins Gilman asked the same question of President Theodore Roosevelt. In the January 1909 issue of *Good Housekeeping* Gilman decries the fact that no women were appointed to the president's Country Life Commission, even though one of the commission's goals was to find ways "'to brighten home life in the country and make it richer and more attractive for the mothers, wives and daughters of farmers'" (120). Gilman points out that neglecting to include women on the commission feeds the idea that only men are farmers, leaving farm women to be "lumped together as mere feminine connections of men" and to be "studied into and recommended about as if they were part of the live stock" (120).

The following month, in apparent response to Gilman, *Good Housekeeping* published Professor Charles W. Burkett's "What the Farm Home Needs." A typical view of the position of women in agriculture, Burkett's piece acknowledges farm women's problems but refuses to grant women status as farmers. Instead, Burkett calls for "a new standard of home living, a new arrangement of home and house things that shall provide for this noble queen with a dignity and appreciableness more in keeping with her nature and demands" (149). Industrializing the home will create this new standard and new arrangement, Burkett argues; women must adopt external, technological solutions to make themselves happier: "gentle reader, you must initiate these things yourself. Your husband may mean well, but oh, he is so slow!" (149). Laborsaving changes were, of course, welcome, but far from liberating women from their positions as "mere feminine connections of men," such changes reinforced them (Gilman 120). Burkett's solutions kept farm women invisible by confining them to the house:

> All of this means that the woman herself must have a clear notion of her work. She needs to know how to arrange the kitchen, how to set the table, what to provide for in the living room, how to make the home cozy and comfortable. More than this, she needs to know how to get the best effect in arranging her hair and clothing; what is good taste in music and pictures; and, in addition, she needs to know a good deal about the real sciences of cooking and household work. Here is where the real art of housekeeping and of happiness begins in the home. (150)

As they always have, farm women do tremendous amounts of work, but until recently, little was known outside the family about the work they actually do (Feldman and Welsh 30). Despite the late attention among rural sociologists and historians, farm women's labors still remain "invisible and uncounted" (Leckie 309, 311; see Sachs 11). As the *Post* example suggests, the media simply perpetuates this cultural blindness. To be visible is to be valued, but working as they do within a male and urban-defined agriculture, farm women often go unseen and unappreciated. Abetting their invisibility is the agrarian myth, which defines the yeoman farmer exclusively as male: " 'Yeoman' is gender-specific and implies dominance of both the household and the polity by domestic patriarchs . . . properly used, the term points to power relations that denied female individualism and insisted that women were part of a male-dominated household" (Kulikoff 143–44). In order for the world to have a complete picture of its agriculture, farm women's work must be fully acknowledged. For women to claim an equal place in agriculture, farmwork must be publicly defined for what it is—shared labor of men and women.

Seizing the title of farmer became a real possibility for women with the advent of industrial farming. Because the basis of the New Agriculture lay in profit, systematization, and efficiency, industrial agriculture offered women opportunities to become farmers. Work was done with machinery that could be handled by women, and women, usually the bookkeepers in some farm operations anyway, had new opportunities to move into accounting or management positions for the new, larger farms. With industrial agriculture's stress on intellectual rather than physical work, women could compete with men as farm operators. Unfortunately, the larger culture lagged behind the technology; most people continued to define farm women in Burkett's terms: as housewives, not as farmers. Of course, the New Agriculture itself made sure that farm women would remain invisible: it sold them Burkett's urban, middle-class domestic ideal along with its laborsaving machines.

Early-twentieth-century literature was one place where women imagined women as successful farmers. For example, the central characters of Willa Cather's *O Pioneers!* (1913) and Ellen Glasgow's *Barren Ground* (1925) challenge the usual understanding of farmers as men. But to make themselves visible in a male-ruled world, both characters employ the techniques and assent to the preconceptions of the new farming—hierarchies of work, trust of experts' advice, use of the latest technologies, and domination of nature. Alexandra Bergson and Dorinda Oakley may refute the popular image of the farm woman as farmwife, but each practices to perfection the industrial agriculture described in chapter 1—which is essentially masculine in its codes and outlook. What distinguishes the farming in Norris's *The Octopus* from that in these later novels is that Cather's and Glasgow's industrial farmers are women, not men. Unmindful of agricultural history, literary critics are often content to read *O Pioneers!* and *Barren Ground* as reimaginings of Arcadian landscapes. But removing Alexandra and Dorinda from their agricultural contexts masks the real work they do, thus perpetuating the cultural invisibility of farm women's labors.

## Managing the Farm, Educating the Farmer: *O Pioneers!* and the New Agriculture

Most studies of Willa Cather's *O Pioneers!* comment on Alexandra Bergson's mystic relationship with the land and on the land's positive response to her love, on the "perfect harmony in nature" at the novel's center, or on its country versus city elements (Rosowski 47). In such interpretations, Alexandra is an ideal farmer, one whose literary roots stretch back to Virgil's *Eclogues* (Rosowski 46). Though these readings work perfectly well, they remain incomplete because they ignore a crucial element: the novel's celebration of an agriculture modeled on urban industrialism. Though Cather herself may have had "the dimmest possible view of literature that had a social message," her novel is in fact a demonstration of the early-twentieth-century demand for the New Agriculture, a farming rooted in sound business practices, efficient organization, and scientific discoveries (Woodress 188; Wiebe 126). Advocated by urban agrarians, social scientists, and the U.S. Department of Agriculture, the New Agriculture sought to remake Thomas Jefferson's yeoman into a modern manufacturer, a "New Farmer" (Butterfield 53; Casson 598; see Danbom 23–50).

That its main character is a woman first suggests that *O Pioneers!* challenges Jeffersonian agrarianism and its definition of farmers as men, never as women (Fink 19; Kulikoff 143–44). Defending the national economy as agricultural, Jefferson's agrarianism centered on the small family farm's independent husbandman, a man who was virtuous, hard-working, and faithful to the republic (L. Marx, *Machine* 125–26; H. N. Smith 128–29). But by 1900 the national economy was rapidly industrializing, and though farmers were "commercialists . . . their methods, ideas, and institutions were preindustrial" (Danbom 12). To remedy the latter, the New Agriculture defined the successful farm not as the self-sufficient homestead of agrarian myth but as an efficient, profitable business supporting an increasingly consolidating industrial order.

Though forward-looking in seeking to industrialize agriculture, "urban agrarians were captives of the agrarian myth" (Danbom 34; see Grantham 115). For example, Theodore Roosevelt, who had limited contact with farmers, accepted wholeheartedly Jefferson's agrarianism. The progressive president believed that the farmer "represented the best hope that America had of perpetuating a mighty breed of men"; the farmer was Roosevelt's "last hero as he was Jefferson's first" (Ellsworth 156; see Jellison 2). In his introduction to the Country Life Commission *Report*, the president declared that "the welfare of the whole community depends upon the welfare of the farmer" (10). Like Roosevelt, urban agrarians sincerely believed that Jeffersonian values could be retained as farming industrialized (Bowers 34–35; Danbom 25–28). The Country Life movement as a whole, in urging the New Agriculture, looked backward and forward; it "sought both to preserve traditional agrarian ideals in the face of industrialism and to adapt agriculture to the modern age" (Bowers 134).

## Willa Cather, Urban Agrarian

Willa Cather fits the urban agrarian profile, and not simply because several of her works nostalgically imagine rural Nebraska (Woodress 243, 299).[1] Her early adult life follows the contours of the Progressive Era, 1890–1917; as Guy Reynolds points out, "as a life it is an almost archetypal progressive success story" (13). Raised in rural Nebraska, Cather graduated from the state university in 1895, wrote for several regional newspapers, taught in Pittsburgh high schools beginning in 1901, traveled widely in Europe, and published two books before landing a job in 1906 on the leading progressive journal of the time, *McClure's* (Gerber 36, 40–41, 49). As a close observer of her home state and as someone whose *McClure's* work kept her abreast of the major intellectual streams of her day, Cather was surely aware of the transformation of American agriculture. Robert W. Cherny points out that "when Cather returned for her occasional visits [to Nebraska], she could not have missed the outward signs of [farmers'] prosperity. The pioneers' soddies gave way to substantial frame houses and barns" ("Nebraska" 235). In "Nebraska: The End of the First Cycle" Cather recalls the "rapid industrial development of Nebraska, which . . . was arrested in the years 1893–97 by a succession of crop failures and by the financial depression which spread over the whole country. . . . The slack farmer moved on" (238).

While Cather was managing editor of *McClure's*, the New Agriculture was attracting wide attention in popular national magazines such as *Outlook*, *Independent*, and *World's Work* (Woodress 190; see Danbom 186). *Outlook* devoted its 10 April 1909 number to Country Life issues; in describing the future of agriculture, Charles Dillon argues that "farming in the next generation or so will be more and more scientific. . . . Farmers will live in towns or cities and go to their fields as a business, just as any business man or skilled laborer now goes to his work" (831). Similarly, in the October 1912 issue of the *Atlantic Monthly*, Roy Hinman Holmes declares that "the new farming is of necessity a specialized department of urban life" (523). Holmes defines agriculture as "a form of manufacturing, and its development must be along the lines marked out by the development of manufacturing in the past" (522). Herbert N. Casson asserts, in "The New American Farmer" in the May 1908 number of *Review of Reviews*, that "the new farmer . . . is a commercialist,—a man of the twentieth century. He works as hard as the old farmer did, but in a higher way. He uses the four M's—mind, money, machinery, and muscle; but as little of the latter as possible" (598).

In its representation of the New Agriculture, *O Pioneers!* argues that a successful agriculture works by the same principles underlying twentieth-century industry: market speculation, a hierarchical division of labor, shrewd management, and the continual deployment of the latest technologies. In social terms the text imagines farming as a business, not a way of life. In political terms it answers agrarian radical movements, such as populism, that saw farmers as potentially independent and self-sufficient if it were not for monopolistic urban industrial forces. In keeping with the progressive spirit of the age, the text represents the New Agriculture as a firm belief in better days to come, if only farmers applied industrial organization to control the biological processes of farming. But the novel's most obvious move is still its most radical: a female farmer at its center, *O Pioneers!* models how well a woman can work the new farm.[2]

## "Up and Coming on the Divide, eh, Alexandra?"

Pointing to the rural progressive, agrarian/industrial tension it embodies, *O Pioneers!* opens on a harsh, gray day in January, a month for looking back and looking ahead. Suggesting the dire consequences of holding to a pre-industrial agriculture, Part One, "The Wild Land," begins by imagining a sense of precariousness: Hanover "was trying not to be blown away . . . dwelling-houses were set about haphazard on the tough prairie sod; some of them looked as if they had been moved in overnight, and others as if they were straying off by themselves. . . . None of them had any appearance of permanence" (11).[3] To settle the country and to achieve the ordered community of Part Two, "Neighboring Fields," pioneer John Bergson must die, clearing the way for Alexandra, a young woman who looks "into the future" (21). As the first chapter ends, we see her riding home holding a lantern, a beacon "held firmly between her feet, . . . [its] moving point of light . . . going deeper and deeper into the dark country" (24). At the same time, to make room for progressive farmers, mossbacks must be removed: young Carl Linstrum, who "seemed already to be looking into the past," soon abandons the Divide with a father who "was never meant for a farmer," a man who sells out to Alexandra (21, 52, 65).

The Bergsons advance from struggling as immigrants working their own homestead to becoming "rich as barons" by speculating on neighbors' land (105). To be successful, Alexandra knows early that she must adopt the tactics of the "men in town who are buying up other people's land," men who "don't try to farm it," speculators (66). Like these men, she "read[s] the papers and follow[s] the markets" (28). She emulates real estate man Charley Fuller, who is "feathering his nest" by buying land on the Divide; Alexandra allies herself with Fuller's economic sense of things when she comments, "If only poor people could learn a little from rich people!" (58–59). In sixteen years, she does learn; she moves from working one farm to managing several because her speculative business skill separates her from "bad farmers, like poor Mr. Linstrum" (59).[4]

The text depicts speculation as the breeder of success; playing the market stimulates an economic blossoming that creates an organized landscape. Part One ends with Alexandra convincing her brothers to speculate, to take a "big chance" by mortgaging their homestead to buy up other farms (63). Alexandra has a "new consciousness of the country, felt almost a new relation to it . . . she felt the future stirring" (68–69); when Part Two opens we immediately see that her gamble has paid off. Even the Divide itself is transformed; the former "wild land" (26) is now "squares of wheat and corn" (73). Whereas before the "record of the plow was insignificant" (25), sixteen years later "the brown earth . . . yields itself eagerly to the

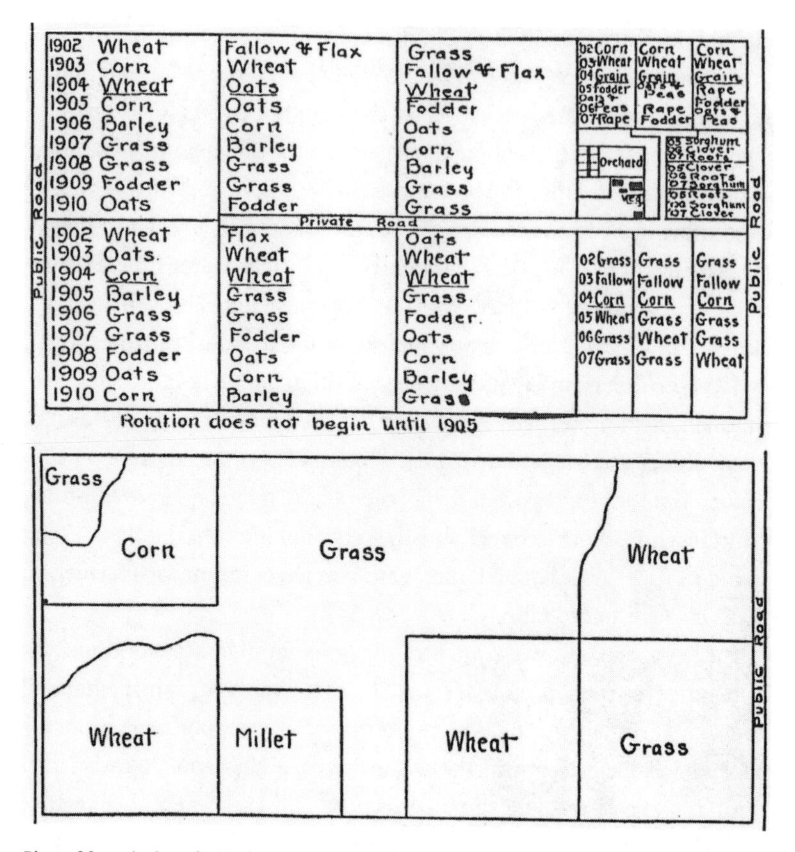


Plan of farm before (*below*) and after (*above*) laying out into regular fields and plan for systematic rotation of crops. From A. M. Teneyck, "Farm Management," in *Cyclopedia of American Agriculture*, edited by L. H. Bailey (New York: Macmillan, 1907), 1:92.

plow" (74). Marks of technological success fill the Divide: "[t]elephone wires," "gilded weather-vanes," "steel windmills" (73). "Order and fine arrangement [are] manifest all over the great farm; in the fencing and hedging, in the windbreaks and sheds, in the symmetrical pasture ponds" (81). Whereas before "the homesteads were few and far apart; here and there a windmill gaunt against the sky, a sod house crouching in a hollow," now "from the graveyard gate one can count a dozen gayly painted farmhouses" (21, 73). Just as twentieth-century industrialism masks the connection between labor and labor's product, the text masks the Bergsons' work by skipping the sixteen years when their hardest labors are expended. Readers see only the result, the bottom line, a profitable landscape.

Readers also see a prominent woman farmer, not a homebound farmwife. Maintaining an active social life—visiting, corresponding, hosting parties—Alexandra is out and about, busy negotiating deals with other farmers, handling bankers, and building her own house (91, 108). Perhaps most significantly, however, she journeys freely beyond the farm. A major complaint of many early-twentieth-century farm women was that they were stuck at home. Unlike many of her real-life contemporaries, Alexandra takes long trips, she does not have to ask a man for permission to go, and her work gets done while she is gone. For example, with her oldest brothers at home, she spends five days studying the river country to be sure that her faith in the Divide is well placed (63). She also travels—unescorted—to Lincoln to visit Frank Shabata in the state penitentiary (255). At novel's end, she suggests to Carl a trip to the goldfields of Alaska (271). Often outbound, this new farmer is quite visible, near and far.

Not all women in the novel are so visible—or mobile. Oscar's wife, for example, we never see; she is confined to her house because she is pregnant. Annie Lee, Lou's wife, briefly appears at a family picnic. Meeting the returning Carl for the first time, she is impressed by his "urban appearance" (103). And Annie makes sure that Carl knows that she will not be a farmer's wife for long; she claims that her family will soon "move into town" and that "Lou is going into business" (104). But her husband dismisses her: "That's what she says" (104). Popping us out of the story for a moment, the narrator stresses how representative the invisible Annie is: "Young farmers seldom address their wives by name. It is always 'you,' or 'she'" (104). Tellingly, Lou always addresses his sister by name. He knows who runs the show.

But the promise of visibility that *O Pioneers!* offered farm women did not become reality. Even as the novel was being written, urban agrarians were remaking farm women into urban middle-class housewives (Jellison 3). With the passage of the 1914 Smith-Lever Act, a significant Country Life legislative achievement, a distinction between men's and women's farmwork became official U.S. policy (Fite, *Cotton Fields*, 81). Smith-Lever was passed, in part, in response to the urban agrarian argument that the rural exodus to cities was "*largely a woman movement*" (Quick 427; Jellison 3). Country Lifers believed that if isolated farm women had access to urban, middle-class "fruits of progress in the house," they would be less likely to convince their families to leave the farm (Quick 429). Rural reformers encouraged the mechanization of housework, and female home economics agents taught farm women efficient homemaking (Jellison 33).[5]

Reformers essentially drove farm women into the house, to appear again only when their husbands called them to the barn or the fields. Establishing the Agricultural and Home Economics Extension System, Smith-Lever "promoted the idea of separate spheres on American farms, with men's work taking place out of doors and women's work being performed in the house"—an idea Alexandra herself accepts (Jellison 16). This is perhaps the novel's most significant irony because Smith-Lever did not simply sanction a preexistent reality; instead, it created a new one: "On the family farm, there were no separate spheres for women and men" (Neth 17; see Jellison 36). The novel thus contributes to the problem it seeks to undo—the invisibility of farm women.

This irony is starkly illustrated in the marriage of Signa, Alexandra's serving girl, to Nelse, her field hand. Hiring recent immigrant labor, Alexandra loses serving girls like Signa as fast as they arrive: "As soon as I get the girls broken in, I marry them off" (204). If the Signa-Nelse marriage is representative, the unions are not egalitarian matches; as she leaves for her new home on her wedding night, the "perplexed" Signa tells Alexandra, "I ta-ank I better do yust like he say" (203). Alexandra, however, is not too concerned about the new bride's future; she simply notes that most of her serving girls "married men they were afraid of" (204). She even comments, if only wryly, that Swedish women all believe that a "cross man makes a good manager" (204). Alexandra condones a type of marriage that she herself avoids. Just as Nelse will remain a hired hand, Signa will remain a serving girl, though now her boss will be her husband. It is hard to imagine the giggling Signa with the "gloomy" Nelse living a life as independent and free as Alexandra's (85).

Industrial farming's demand for separate spheres and labor efficiency is also represented in the Bergsons' early specialization of work. Before their father's death, Alexandra and her brothers shared field labor, but as John Bergson lies dying a distinction between house and field takes shape. John initiates this specialization when he tells his sons: "Alexandra must not work in the fields any more" (32). He realizes that she makes more money selling eggs and butter than she does laboring as a field hand. Following his death, Alexandra works only around the house, tending her chickens, making butter, but more importantly, doing the farm's planning and accounting, while her brothers toil in the fields—under her orders. Labor and management cease to be one and the same, as they had been in the figure of John Bergson. After the family subdues and organizes the landscape, the two brothers marry and the farm is divided among the three siblings. Soon the sole proprietor of several farms, Alexandra makes finer distinctions regarding work by reordering into separate spheres household chores, business management, and farm labor. While she spends her time working with her farms' accounts, serving girls do household chores and hired men do fieldwork. She later promotes herself to a higher management position when she hires Barney Flinn as a "foreman" to manage her farms' laborers—though, like a factory boss, she might be seen "overseeing the branding of the cattle, or the loading of the pigs" (85, 186). This management-labor hierarchy allows Alexandra to secure the settlement of the "wild land," ultimately making herself and her brothers "independent landowners, not struggling farmers" (65).

Reflecting the New Agriculture's view of farming as a business rather than as a way of life, *O Pioneers!* argues that farmwork is first white-collar work. The text legitimizes management as work most directly in Alexandra's confrontation with Oscar and Lou over her involvement with Carl. The brothers interrupt her when she is "busy with her account-books" to insist that they have claim to Alexandra's land because they have done the physical labor that has made it prosper (149). Alexandra replies by distinguishing between physical work and mental work (management). Claiming that her work "puts in the crop, and it sometimes keeps the fields for corn to grow in," Alexandra defies her brothers' natural rights logic (153; Horwitz 222). Her reminder that her brothers were always ready to give up their labors and that they always balked at each of her experiments underscores the novel's point that farm success derives mainly from persistent and sound management, not from heavy and consistent labor (153). John Bergson himself realizes that his sons have no understanding of farm management: "Lou and Oscar were industrious, but he could never teach them to use their heads about their work" (28).

## The Brain-Working Farmer

The Bergson family represents a triadic view of the farmer shared by many Country Lifers. Rural sociologist Kenyon Butterfield claims that "the old farmer was in his day a new farmer; he was 'up with the times.' . . . The new farmer is merely the worthy son of a noble sire; he is the modern embodiment of the old farmer's progressiveness. The mossback is the man who tries to use the old methods under the new conditions" (53). Alexandra's father, John Bergson, is an old farmer, the intelligent pioneer. As Neil

Gustafson argues, John is no "failed farmer"; in fact, he dies bequeathing "'his hard-won land,'" and he and Alexandra have a "shared dream" of the Divide (152–53). In the view of the New Agriculture, the failures are Oscar and Lou. Oscar is clearly imagined as a mossback: he "liked to begin his corn-planting at the same time every year, whether the season were backward or forward. He seemed to feel that by his own irreproachable regularity he would clear himself of blame and reprove the weather" (56). The brothers are men who "were meant to follow in paths already marked out for them, not to break trails in a new country. A steady job, a few holidays, nothing to think about, and they would have been very happy" (49–50).[6]

This triadic view of the farmer dovetails with the novel's representation of the New Agriculture's emphasis on brainpower over physical power, an emphasis that recalls and extends similar representations found in William Allen White's bonanza article and Norris's *The Octopus*. *O Pioneers!* describes a rural society in which farmers employ "methods of farming requiring the highest intelligence"; the "brain-working farmer is the man behind prosperity" (*Report* 30; Casson 599; see Collins 18–19). In the early twentieth century, "Machinery, technology, and scientific methods changed farming from manual labor to intellectual labor" (Neth 218). The novel stresses this when Alexandra criticizes "these stupid fellows," those leaving the Divide: "Why are we better fixed than any of our neighbors? Because father had more brains. Our people were better people than these in the old country. We *ought* to do more than they do, and see further ahead" (66–67). The intensity of the text's negative portrait of Oscar underscores how much it values intellectual work over physical labor: "He was a man of powerful body and unusual endurance; the sort of man you could attach to a cornsheller as you would an engine. . . . His love of routine amounted to a vice. He worked like an insect, always doing the same thing over in the same way, regardless of whether it was best or no. He felt that there was a sovereign virtue in mere bodily toil, and he rather liked to do things in the hardest way" (56).

The brothers do not share their sister's quality of mind: whereas Lou is "apt to go off at half-cock" and Oscar is "indolent of mind" (55–56), Alexandra is intelligent "like her grandfather," a successful shipbuilder, a man who "built up a proud little business with no capital but his own skill and foresight" (28–29). In the midst of "The Wild Land," we constantly see Alexandra thinking, planning, gathering information and advice—using, unlike her brothers, her head about her work (28). Like her "powerful" grandfather, she has "the strength of will, and the simple direct way of thinking things out, that had characterized [him] in his better days" (29). Recognizing his daughter's mental superiority over her brothers, John Bergson leaves his farm in Alexandra's "strong" hands (30). At this key transitional moment, intelligent farming is imagined as the act of strength that creates agricultural success, a view shared by proponents of the New Agriculture: "weaker farmers will be unable to sustain themselves; the weaker farmers will be those who direct their labor least wisely; these again will be those who know least" (Adams 21).

## Prudent Fertility

As a middle ground between wilderness and mining the soil, Alexandra's farm demonstrates rural progressivism's marrying of agriculture and conservation (Bailey 198–200). In the Progressive Era, during Roosevelt's administration especially, "the conservation and country-life movements rest[ed] on the same premise"; for urban agrarians, this meant "utilizing the products and forces of the planet wisely" (Bailey 179; Bowers 38). Alexandra mediates between Crazy Ivar's and her brothers' land uses by establishing a profitable farm that systematizes nature. Ivar "lost his land through mismanagement" because he kept his farm wild (83). Representing a way of dealing with nature at odds with the brothers' exploitation and Alexandra's skillful management, Ivar lives without disturbing the land, a mark of inefficient land use for a New Agriculture stressing "a system of diversified and rotation farming" (*Report* 90): "Ivar had lived for three years in the clay bank, without defiling the face of nature any more than the coyote that had lived there before him had done. . . . He preferred the cleanness and tidiness of the wild sod" (39–41). Oscar and Lou, who disdain Ivar since he will never "be able to prove up on his land because he worked it so little," exploit nature (47). As boys, they shoot birds for fun; as self-satisfied adults, they take cherries from Alexandra's orchard because they have no "patience to grow an orchard of their own" (98). In their selfishness, the brothers represent the nineteenth century's careless, "primitive system of land exploitation" which the New Agriculture meant to replace with order and precision (*Report* 84; see Davenport 45–46). Touted as the nation's chief soil conservator (Bowers 38), the new farmer reconciled John Muir's spiritual preservationism and Gifford Pinchot's utilitarianism (Bates 81; Unger 647).[7]

Alexandra accomplishes a twin urban agrarian goal: "both Emil and the country had become what she had hoped. Out of her father's children there was one who was fit to cope with the world, who had not been tied to the plow, and who had a personality apart from the soil" (191). A Country Life success story, Emil grows from a "clumsy" country boy to a university graduate who "can scarcely remember" his sister's struggle with the "old wild country" (12, 76, 75). The son of Swedish immigrants, Emil is "just like an American boy, —he graduated from the State University in June" (108). His transformation at the university parallels the Divide's transformation following the implementation of university ideas, such as alfalfa. By twenty-one, Emil has become "the best" there is on the Divide, a man full of possibility (271). Through her years of struggle, Alexandra has mothered him to give him "a chance, a whole chance" so that he can "do whatever he wants to" with his life (109). Having lived in Mexico City, the adult Emil stands ready to fulfill the Country Life vision of the farm supplying "the city and metropolis with fresh blood, clean bodies and clear brains that can endure the strain of modern urban life" (*Report* 31; Danbom 25). But his promise goes unfulfilled because unlike the patient, prudent, cost-accounting Alexandra, he lives "at the mercy of storms" and is incapable of intelligently managing his passions (202; see 162). Allowing himself to be "overtaxed by excitement and sorrow" (228), and instead of moving on to Omaha and law school, he returns to the Shabatas' orchard, Marie's "neglected wilderness" (138). A misplaced nostalgia kills "bad boy" Emil and his "impulsive" lover, Marie (253).

## "Her Training Had All Been toward the End of Making Her Proficient in What She Had Undertaken to Do"

Alexandra's farming abilities are not as innate as many critics suggest; they are acquired and disciplined. In reworking her homestead's wild land, she is guided by land grant universities, which were created in the mid-nineteenth century to serve as resources for American agriculture. Boosters of the New Agriculture urged farmers to utilize university advances in scientific agriculture, something most farmers were reluctant to do (Danbom 88; Bowers 111; Neth 104; Adams 28). To combat this hesitancy, the Country Life Commission advocated the creation of "a well-developed plan of extension teaching conducted by agricultural colleges, by means of the printed page, face-to-face talks, and demonstration or object lessons"

(*Report* 26–27). *O Pioneers!* promotes the Country Life insistence that university experts should guide advances in agriculture; new ideas should no longer be farm-grown as they once had been. Several times the novel points to the positive results of Alexandra's access to the University of Nebraska. For example, she learns about a "new kind of clover hay" (63) from a "young man who had been to the University" (154). The Cornhusker-educated Emil is an intellectual resource for her; it is hinted that Emil, "with his university ideas . . . instigated the silo" (86). Before she visits Frank at the state penitentiary, Alexandra strolls by the University of Nebraska, feeling a "great tenderness" for the male students who "come running down the flagged walk and dash out into the street as if [they] were rushing to announce some wonder to the world" (256). The one university student she talks to makes her feel "unreasonably comforted" in her grief over Emil's death (257).[8]

Though agricultural education is embodied favorably in Emil, its absolute necessity is represented in Oscar and Lou's pathological suspicion of Alexandra's technological experiments—until she demonstrates their feasibility. Her brothers must see the viability of land speculation, wheat, alfalfa, and silos before they will adopt them. Understanding that many farmers were like Oscar and Lou, banks and businesses dependent on farming underwrote demonstration farms to "promote agricultural efficiency and prosperity" (Bowers 89). Precursors to the county extension system, these demonstrations were the "best solution available for the problem of adult education in agriculture" because demonstration agents had "the ability to supervise farmers and to follow up on instruction" (Danbom 71). Agents taught not only scientific farming but also " 'economy, order, sanitation, patriotism, and a score of other wholesome lessons' " (Danbom 72). Itself a demonstration of progressive farming, *O Pioneers!* represents for its primarily urban audience the need for "the new generation of scientific farmers" to redefine the nation's agriculture through "redirected education" (Foght 149; *Report* 121; Butterfield 79; Bowers 4–5; Danbom 55–56).[9]

Getting farmers to adopt industrial technologies was a key component of the New Agriculture: "the mastery of machinery—the transformation of the farm into a factory . . . gives [the modern farmer] a sense of mental superiority never before found upon the farm" (Collins 316). Farmers who refused to adopt new technology were labeled " 'backward' " (Neth 97). *O Pioneers!* valorizes Alexandra as someone unafraid of new technology.

While exploring the river country farms with Emil, she "spent a whole day with one young farmer who had been away at school, and who was experimenting with a new kind of clover hay" (63). This hay helps to replace the wild land's "shaggy coat" with a "vast checker-board" of neighboring fields (73). Clover gives way to Alexandra's successful experiment with alfalfa—"the salvation of this country"—a perennial introduced to Nebraska during the period when "Nebraska agriculture may be said to have come into its own . . . between 1890 and 1908" (154; R. P. Crawford 98). In the face of Lou's resistance—"everybody [is] laughing at us"—Alexandra puts in "the first big wheat-planting," a practice her neighbors adopt only after seeing her "three big wheat crops" (154). In addition, she has built the "first silo on the Divide," and, though her hired hands criticize her for it, we know that her experiment will succeed (85).

Adopting new technology is fine, but the progressive farmer needs to know how to use it properly. Alexandra's neighbor Amédée Chevalier also runs a highly mechanized farm; he operates a steam thresher and a header (215). But, unlike Alexandra, he has inexpertly managed his purchases. He is the only one who can run both pieces of equipment, so "he has to be everywhere at once" (215). His precipitous investment—"three thousand dollars' worth of new machinery to manage"—keeps him in the field when he should be in the hospital (218). He is "overheating himself" physically and economically because he has not made Alexandra's split between labor and management (216). Clearly, he is worried about whether his crop will pay for the technical improvements he purchased to harvest it. His "wheat's short" and ready "to shatter" (218), and paying off his investment rests on his wife's "hope he can rent it out to the neighbors, it cost so much" (215). The stress of all this contributes as much to his death as his appendicitis. In his last act in the field, Amédée is waving "to the engineer not to stop the engine" (218).[10]

The machine in this novel's garden is a positive, creative force, not the interrupter of a rural pastoral moment.[11] It is not the machinery that kills the happy Amédée; his mismanagement of it kills him. Across the Divide positive images of the machine abound: telephone wires "hum," the land "yields itself eagerly to the plow . . . with a soft, deep sigh of happiness" and "the grain . . . bends toward the blade and cuts like velvet" (73–74). These machines make the "order and fine arrangement" of Alexandra's farm (81). Even the rifle, the novel's most insidious machine, serves the New Agriculture's efforts to efficiently control natural processes and spontaneity by killing off the novel's impetuous lovers, Emil and Marie, and by returning our and the novel's attention to the text's ordered heroine, Alexandra.

## "But It Gratified Him to Feel Like a Desperate Man"

To transform rural society, the New Agriculture needed to contain lingering political passions of agrarian radicalism. *O Pioneers!*'s vision of a New Agriculture culminates in its dim view of populism, an agrarian extremism that severely critiqued industrial capitalism (McMath 111). Ethnographer-anthropologist Deborah Fink notes that "twentieth-century reformers did not like what they actually saw in the countryside—particularly the Populists. They feared rural agitation" (25). The New Agriculture marked a move away from not only nineteenth-century farming but also its unsettled politics: "Where of old [the farmer] spent the long evenings brooding over fancied wrongs and came to believe himself a victim of machinations and of circumstances, now he goes out and helps to manage and is part of the industrial world" (Harger 843). In contrast to the shrewd Alexandra, Lou is a William Jennings Bryan backer, a Populist "political agitator," who mismanages his farm (104, 136). Unable to make as much money as even Oscar, Lou is "tricky . . . he has not a fox's face for nothing . . . he neglects his farm to attend conventions" (93). Lou himself unwittingly links Populist convention-going with insanity when he complains to Alexandra about Ivar: "When I was in Hastings to attend the convention . . . I saw the superintendent of the asylum" (94).[12]

While "Prudent Alexandra" invests in technology to improve her farms, Lou spends his money extravagantly; he indulges his wife's preoccupation with "rings and chains and 'beauty pins'" and buys a bathtub that Annie declares is "weakening" him because he stays in too long (267, 93, 96).[13] More significantly, Lou is jealous of eastern establishment money: "We gave Wall Street a scare in ninety-six. . . . Silver was n't the only issue. . . . The West is going to make itself heard. . . . We have a good deal more to say than we had when we were poor. . . . We're getting on to a whole lot of things" (104–5). But his politics are violent. He encourages Carl Linstrum and other "folks in New York" to "march down to Wall Street and blow it up. Dynamite it, I mean" (104–5). Though Lou's populism may menace a capitalist ideology, his threats are futile: the urban Carl recognizes that the "same business would go on in another street. The street does n't matter" (105). Radicals like Lou, or Frank Shabata, cannot stop the impending marriage of metropolitan New York and rural Nebraska. As if answering Lou's extremism, new farmer Alexandra and gold prospector Carl marry—as friends, thus equal partners, in order to be "safe" (273).[14]

Allied to Lou is the jealous Frank, the county's other political agitator and a murderer. Every Sunday he decries the excesses of the Gould family by telling an "inexhaustible stock of stories about their crimes and follies, how they bribed the courts and shot down their butlers with impunity" (136). Marie hates to see the newspapers come because she "had nothing but good will" for the Goulds (135). Frank is as jealous of the Goulds' money as he is of his wife's affections: "If he ever got rich he meant to buy her pretty clothes and take her to California in a Pullman car . . . in the mean time he wanted her to feel that life was as ugly and as unjust as he felt it" (238). An anti-Populist portrait of agrarian extremism, Frank is "a desperate man" whose "unhappy temperament was like a cage," a man who "made his own unhappiness" (234). Murdering Marie and Emil is his most radical and futile gesture at the forces he imagines arrayed against him. Not content to imprison him, Cather wants him exiled. If the progressive Alexandra can get him pardoned, the Populist Frank declares that he will "not trouble dis country no more" (263).[15]

Willa Cather's *O Pioneers!* presents us with a successful agrarian heroine of almost mythic proportion who models her farming on urban industrialism to transform an unproductive land into a lush breadbasket. In Alexandra we see the best demonstration of two things: the viability of the New Agriculture and the ability of women to practice it. Foreshadowing today's agribusiness, the text praises market speculation, technological change, and hierarchical farm labor divisions. In picturing farm life positively, *O Pioneers!* answers grim portraits of farm life drawn by Populists such as Hamlin Garland and envisions an agriculture that will sustain the expansion of urban American industrialism with cheap food and displaced labor.

## What the Farmer Really Looks Like

*O Pioneers!* and Ellen Glasgow's *Barren Ground* (1925) appeared in a context studded with stories that assumed a connection between women's emancipation and progressive farming. For example, in "The Man from the City" George Wayne, an urban muckraker of "the sorrows of the cities," takes his doctor's advice to vacation in the country, "where a real thought is as rare as a pterodactyl" (Hay 233). He boards with the Millwoods, a farm family marred by domestic violence. Wayne's "quick, flamelike" movements contrast sharply with the unprogressive Harry Millwood's "dragging heaviness" and his reputation for being "lazy" (234, 236, 240). Wayne quickly warms, however, to the efficient and resourceful Mrs. Millwood, "a woman whose every hour was heroism . . . [who] slaved" while Harry "gossiped at the store and left his wheat to rot in the fields" (241). She chides her husband for delaying the threshing, pointing out that the region's most progressive farmer, the "well-to-do," moneymaking Tom Thornton, has already threshed his (236). But Harry dismisses her, replying, "Thornton's no pattern for me" (239). When Harry later refuses to seek medical attention for his son Richard, an enraged Wayne diagnoses the boy with appendicitis, forcefully borrows a neighbor's car, fetches the local doctor, and brings in his own urban doctor to save the child's life. In the story's world, country life would be better materially and socially if women like Mrs. Millwood were freed from mossbacks like Harry; national life would be better if rural Mrs. Millwood could marry urban Wayne.

A second example, "An Early Spring," tells the story of Mrs. Taggart, a small-farm apple grower who is certain that her trees are dying. At fifty, after studying voice in Europe and being "cruelly deceived" in love, she has borrowed money to purchase a Pennsylvania farm, asking only for "peace and a livelihood from this land" (Singmaster 206). After a killing frost, five male neighbors appear at her door to congratulate her on her "good luck," claiming that she was "smarter" than they were because she had planted the "latest variety"; now the "envy of [them] all," the independent Mrs. Taggart will harvest "the only crop in the county . . . to sell at [her] own price" (211–12). With her fortune secure, she plans to buy a car to replace her dead horse, to "give orders for the next spraying," and to help the "little owners to whom the blow would be severe" (212). Like Alexandra and Dorinda, she proves that she can farm as well as any man.

But in the 1920s the dominant portrait of the farmer—whether progressive or not—was still male. Responding to the cinematic vision of the farmer as a hayseed with a "scrawny beard, high boots, [and] tucked-in trousers," the *Country Gentleman* began a series of cartoons in July 1921 that sought to find out "What a Farmer Really Looks Like" (Tilden 7). Cartoons were solicited nationwide and the series ran for several months, depicting the farmer as a bewhiskered old man exchanging a handful of money for urban gold (Marcus); Uncle Sam shooting down "pessimism" (Sykes); a man in a tie beating off bugs, blight, and bad weather (Rohn); a businessman in suit and tie (Chapin; G. Williams). The cartoons illustrate a tension between the expected urban stereotype and the reality, between the old farmer and the new, progressive farmer.

Though several cartoons satirize the standard image of the farmer as a graybeard in overalls and slouch hat, they still depict him so. Most of the cartoonists realized, as an *Omaha World-Herald* cartoonist puts it, "A cartoon farmer without the alfalfa does not carry conviction" (Spencer 13). The public expected the old image and got it, even though, as a *Chicago Daily News* cartoonist claims, the " 'rube' farmer [has] simply ceased to exist. Nothing but city people in old U.S. now; only a lot of them refuse to live in the cities" (T. Brown 4). Answering those cartoonists who depict the farmer as a businessman in suit and tie, *Indianapolis News* cartoonist Gaar Williams argues that his paper will still "make pictures of farmers that look like dirt farmers in honor of the farmers who still look like dirt farmers" (5). But alongside his whiskered "old farmer model" Williams pictures the "new farmer model"—a man in suit and tie who claims to be a "real 'dirt farmer' " (5).

Targeting farm readers in the *Country Gentleman*, these urban-created cartoons assured farmers that urbanites understood that they were modern and progressive, not backward and unsophisticated hicks. At the same time, the cartoons suggested to farmers who were not progressive that they should be. And yet, though these cartoons redrew the farmer, they did not escape the assumption that farmers are men. No women appear in the cartoons' definitions of "what the farmer really looks like" (Tilden 7). The assumption that only men were farmers and that farm women were housewives was firmly in place when Ellen Glasgow's *Barren Ground* appeared in 1925.

## "Passion Transfigured": *Barren Ground* and the New Agriculture

Critics have often read *Barren Ground* through the pastoral prism of the agrarian myth. For example, Elizabeth Harrison claims that with *Barren Ground* Ellen Glasgow "transforms the pastoral myth from a male- to a female-centered quest for heroism" and that the novel's main female character, Dorinda, "embodies the author's yeoman farmer ideal" (29, 33). In a similar vein, Jan Zlotnik Schmidt argues that the work ethics of Glasgow's heroines are "legends of ordered pastoral worlds, Glasgow's reinterpretations of the American dream (the self-sufficient American living off the soil)" (119). Tonette Bond states the case for this argument succinctly: "The story of Dorinda Oakley is the record of her redemption through the reordering of her physical and mental environments to accord with the pastoral ideal" (565).

But, like Cather's *O Pioneers!*, Glasgow's novel defines farming in industrial terms. The text promotes a farm economy built on market speculation, rapid technological change, strict divisions of labor, and dependence on university experts. In contrast, most farmers defended agriculture as a way of life; they conflated management and labor in the farmer, practiced proven rather than experimental farm methods, assigned work and property intrinsic value aside from their exchange value, and looked to the local community as the best source of agricultural knowledge. Dorinda Oakley's rise from daughter of a "'land poor'" farmer to an independent agribusinesswoman managing two reclaimed farms suggests that *Barren Ground* champions an urban insistence on the application of industrial techniques to agriculture (7).[16]

In framing the novel's time and place, the narrator situates *Barren Ground* squarely within this agricultural debate, noting, "Thirty years ago, modern methods of farming, even methods that were modern in the benighted eighteen-nineties, had not penetrated to this thinly settled part of Virginia" (4). Contemporary reviews paid much attention to the novel's farm focus: "Back to the Soil" (Boynton), "Soil and Soul" (Henderson 265), or "Down on the Farm" (Languish). One reviewer argues that "Dorinda's long struggle with the land somehow resembles a success story from an agricultural magazine" (Green 119).[17]

The pervasiveness of such readings prompted Glasgow to counter that her novel was not concerned with "systems of agriculture" (*Certain Measure* 160). But defining farm practices as systematic only underscores her assumption that agriculture can, in fact, be systematized, a belief touted by agricultural reformers but one not held by all farmers. Unfortunately, many recent critics simply follow Glasgow's lead, asserting that the text is "responsive to Southern agrarian sentiment" and that Dorinda is "the mythic, golden-age laborer, kin to and wedded to the soil, affirming the old agrarian values" (McDowell 147; Thiébaux 126). Such conclusions obscure how *Barren Ground* reflects its historical moment's radical redefinition of agriculture, a redefinition that guaranteed the rapid expansion of urban industry and led to the dispossession and displacement of millions (Fite, *American Farmers* 74–77).

Just as *Barren Ground* begins in the 1890s, so do major efforts to redefine the farmer in industrial terms.[18] Rural progressives—primarily urban agrarians, social scientists, and the U.S. Department of Agriculture—argued that the new farming meant "to replace tradition with well established facts; to substitute for the irregular and uncertain purposes of the individual a systematic and well organized business of food production by the community at large" (Danbom 23–50; Davenport 49). Reformers, who "were less interested in the farmers' welfare than they were in cheap food," distinguished between old farmers who adhered to traditional practices and new farmers who were adopting progressive methods and machinery (Fite, *American Farmers* 17). Such reformers smugly defined the old farmer as an unthinking laborer who relied only on muscle to do his work: "the primitive hoe-man's heavy work in the major activities of his occupation cannot automatically . . . awaken an intellectual life" (Galpin, *Life* 33–34). Simply by purchasing machinery, however, the old farmer "becomes a new cerebral type, whose very struggle with the earth summons him to an employment of his hereditary intellectual mechanism, and a consequent intellectual life" (Galpin, *Life* 35–36). Again and again reformers stressed a relationship between progressive farming and intellect: "Rural improvement means developing a better people—more intelligent, more capable of appreciating those finer things that lead to culture . . . the farm can be improved only when the farmer is also improved. . . . Mind culture is essential to proper soil culture" (MacGarr 10; see *Report* 30; Casson 599; Collins 18–19; Fite, *American Farmers* 18).

*Barren Ground* similarly divides farmers into two classes: the old "hoe-farmer" and the new "machine-farmer" (Galpin, *Life* 32–36). Tobacco farmer Joshua Oakley, Dorinda's father, represents the old farmer; his lack of intelligence and his constant physical labor characterize him: he is a "slow-witted man" (40), who, "in spite of his ignorance, had possessed an industry that was tireless" (302). He looms in Dorinda's memory as the "titanic image of the labourer who labours without hope" (348). Eudora Abernethy, his wife, "recoil[s] from her husband's inefficiency" and "despis[es] him with her intelligence" (43). Unlike new farmers who sought greater efficiency and more control over the processes of nature, Joshua is "a slave to the land, harnessed to the elemental forces, struggling inarticulately against the blight of poverty and the barrenness of the soil" (40).

From a machine farmer's perspective, perhaps the most damning description is of Joshua as "a dumb plodding creature who had as little share in the family life as had the horses" (40).

To transform Old Farm from a static homestead to a paying business, the old farmer must die (246). Joshua's simpleminded refusal to rotate crops marks him as a man unable to change things for the better; he tells Dorinda, "I ain't one fur new-fangled ways" (119), and he responds to his son's suggestion that the tobacco beds be moved by saying, "Well, they've always been thar" (55). Ironically, Joshua's decision to make this very change paralyzes him; he has his debilitating stroke while plowing a tobacco field in which "he had decided not to plant tobacco" (264). Underscoring his incapacity is his inability to learn modern farming as Dorinda does; he cannot read or write and lives out his life within Old Farm's bounds (253). At the moment when Dorinda tells "him of the lectures she had heard and the books she had read" in New York, "he lay looking at her with his expression of mute resignation" (267). He dies a farmer with "no language but the language of toil" (118).

In contrast, the articulate Dorinda is an efficient and intelligent farmer who teaches proper grammar to her employees (349). A model new farmer, she invests in new machinery, superintends a division of farm labor, and participates in a specialty cash market. The narrator often remarks on her "intelligence and independence," "her mind . . . crowded with ideas," her "self-reliance" and "calm efficiency" (86, 376, 502). Her brother Rufus comments that she returns from the city "looking as if [she] could run the world" (258). And her mother quietly wonders if it is "possible that she had created this superior intelligence, that she had actually brought this paragon of efficiency into the world" (271). Though she manages her dairy by "keeping a relentless eye on every detail," Dorinda's thoughts are not limited to farming (310). She and John Abner pass their evenings discussing "books and distant countries" (427), a sharp contrast from earlier days when Dorinda's parents gathered before the fire: "Drugged with fatigue, they nodded in a vegetable somnolence" (48).

Many farmers resisted urban-based rural reform (Danbom 85; Neth 146). County extension agents, foot soldiers for the New Agriculture, were the "victims of much derision" and many times met farmer hostility (Danbom 87–88, 132). A book farmer herself, Dorinda prospers despite community ridicule (397). Nathan Pedlar, however, most clearly represents rural progressives whose ideas are derided by farmers. The first time readers meet Nathan—who later dies a hero—the narrator describes him as ahead of his time: "though Dorinda never suspected it . . . [Nathan] had come into the world a quarter of a century too soon" (18). Dorinda defines him as "the only man at Pedlar's Mill who lived in the future," and she is fascinated by his "visionary" ability (421). Despite having "always been ridiculed by his neighbors," he had waged "a protracted battle" to persuade them to use "the telephone, the modern churn, and the separator . . . labour-saving inventions," all of which she adopts herself (421). Nathan even imagines "the time when they would have an electric plant on the farm and all the cows would be milked and the cream separated by electricity. Was this only the fancy of a visionary, or, like so many of Nathan's imaginary devices, would it come true in the end?" (421). Glasgow's contemporaries would have been well aware that Nathan's dream would be realized, thus validating his progressive practicality, the naturalness of technological change, and the inevitability of industrial agriculture. Reminding one of Alexandra's and Carl's, the marriage of book-farmer Dorinda and pragmatic new farmer Nathan represents the ideal union needed for rural uplift.[19]

The choice of butter making as Dorinda's road to success is no accident. Butter making was quite amenable to the rationality, systematization, and control of industrial farming (Galpin, *Life* 49). With the introduction of mechanical cream separators and milking machines, dairy herds expanded; milk was analyzed and classified. To increase production, farmers milked at regular hours, doing chores, as Dorinda does, "not by necessity, as in the old days without system, but by the stroke of the clock. Each milker had her own place, and milked always the same cows" (348). To increase production, farmers also sought more control over cows' reproduction. Registered herds became common, and, like many new farmers, Dorinda is careful to use prize bulls (346). By 1922 rural reformers were noting that "one of the most recent steps" in agriculture's "evolution" has been that "butter making has passed from the domestic into the commercial stage in farm economy" (MacGarr 162). Warren H. Wilson, an early rural sociologist, notes: "Work in the dairy country is not so much seasonal, as it is systematic. The work for the various hours of the day is as rigorously prescribed to the dairy farmer . . . according to the exactions of the city market for milk and the physiological possibilities of the dairy cow" (120). Though these industrial methods and techniques characterize Dorinda's farm, her dairy is first and foremost a female space: her employees are all women working with cows (420).

Unlike *O Pioneers!*, *Barren Ground* is critical of the separate spheres ideal that perpetuated the invisibility of farm women in the New Agriculture. Dorinda's example directly challenges the 1914 Smith-Lever Act. In returning from New York and industrializing the entire farm, Dorinda defies the distinction between men's work and women's work that Smith-Lever had enacted. Readers see her working in the fields (409), in the barn (420), and in the home (429). Refuting the central assumption behind Smith-Lever that farmers are men, never women, Dorinda's skill and persistence eventually "proved that she could farm as well as a man" (387; see Fink 19–21; Kulikoff 143–44; Neth 220). By novel's end, she has rejected several female roles defined by a separate spheres ideology: homemaking mother, homebound farmwife, governess's assistant. Thus the most visible product of her efficient farm practice is not her crocks of butter but her "triumphant independence" (401).

From a Country Life perspective, the most visible sign of her independence is Dorinda's purchase of new technologies; like the Ellgoods, who "went in debt and bought the newest inventions," she has learned how to farm "the right way" (292, 291). Unlike many of her neighbors, she does not fear new ideas. Following the example of the successful Ellgoods, she experiments with "sweet clover with lime" (332). Planting alfalfa, a perennial, becomes "the making of Five Oaks" (428). She experiments with ensilage (346) and electricity (468); her dairy operation is the first in that part of Virginia (291). She invests in the latest machinery—the "separator" (269), a new "tractor-plough" (427), and "electricity" (468), the latter long before most farmers had access to it.[20] The tractor, the main status symbol on the American farm, marked the owner's transition from "farmer as worker to farmer as manager" (Neth 224). Using her car to pick up Jason Greylock at the poorhouse, Dorinda notices that other farmers "stopped in their ploughing or cutting, and turned to stare curiously like slow-witted animals" (497). As Dorinda's mobility and strength visibly increase, her betrayer's visibly weaken; in the end, while Dorinda is known statewide for her skills, Jason survives on her charity, "remote because he had lost all connection with his surroundings" (477, 511).

Whereas in her father's operation of Old Farm labor and management were conflated in him, in Dorinda's operation management and labor are hierarchized. Sensitive to urban radicalism, Country Lifers looked on the farmer as "a third element to balance the violent and avaricious forces of capital and labor" (Danbom 27). A middle-class farmer, Dorinda oversees her workers with "patient discipline" (380); her dairy's "simple tasks came under her watchful eyes" and she "direct[s] the work in the fields" (311, 331). Establishing a division of labor by hiring male field hands and women milkers, she vigilantly keeps everyone "under her supervision" (409). After expanding her holdings, she hires a manager for Five Oaks and names Nathan, her "superior hired man," as Old Farm's manager (387, 409). While male farmers struggle through a post–World War I labor shortage, Dorinda's good reputation among workers ensures that she has enough field hands to work her farms (463). But she is never sentimental about her labor force; to increase profits, she "replaced hand labour by electricity" (468; see 476).

Dorinda's butter marketing contrasts sharply with the community exchange in Pedlar's Mill and the bartering practiced at Nathan's store (76, 81). Participating in a lucrative specialties market, she sells only butter—at a steep price, a price based not on its inherent quality, but on its perceived value (311). She is so deeply embedded in urban sales—her product is sold miles away in the nation's capital—that she forgoes butter and drinks buttermilk "in order that she might keep nothing back from the market" (347). In her skillful hands, Old Farm moves from being an isolated farm with "no milk or butter" to becoming one with a salable surplus tied to urban consumers (41). Ever expanding, the Oakley business booms with the purchase of new land and with the advent of "three trains a day [running] between Washington and the South" (409).

Reformers tied increased income to debt (Neth 219). Country Lifers encouraged farmers to make more use of credit, much the way Dorinda does when she reclaims Old Farm with urban capital, with the speculative investment of a New Yorker, Doctor Faraday. The superiority of long-term investment is a constant in the novel, introduced early through the description of James Ellgood, who inherited a "small fortune" and owns a "flourishing stock farm . . . for five years he had put more into the soil than he had got out of it" (6). But Faraday invests in Dorinda because he sees in her the qualities it takes for future success: her "efficiency" is "remarkable" and she is quite "practical" for someone "without special training" (248). The doctor's networking creates Dorinda's market; his "influence" persuades a Washington hotel to take her butter, making "her success . . . less fortuitous than appeared on the surface" (312). Dorinda borrows more money, because "without the courage to borrow money, she could never have made the farm even a moderate success" (348). With borrowed capital, she improves and expands her holdings and increases her herd's quality (348). She even invests in marriage, virtually by contract; when Nathan proposes, he does so as an investment proposition: "if you could make up your mind to marry me, we might throw the two farms into one" (364).

Though several scholars describe Dorinda's time in New York as a blemish on the novel, it is, in fact, the novel's central episode.[21] In a benign New York Dorinda learns the book farming that changes her life; books do "much good indeed" for her (366). Book farming, learning the theory of agriculture and then practicing it, was resisted by most farmers (Danbom 87–88). When she recognizes the possibilities of agricultural theory, Dorinda rejects the inherited knowledge of Pedlar's Mill: "We have the experience of generations, and it has taught us nothing except to do things the way we've always done them" (243). Like any new farmer, she then turns for advice and direction to university and U.S. Department of Agriculture (USDA) extension experts, experts who believed that "only when a subject has reached the scientific stage . . . it becomes teachable through the elucidation of the principles involved" (Davenport 49). To learn "modern ways of getting the best out of the soil," she reads and listens to the advice of University of Wisconsin experts (242). Later she admits that she "found out all [she] could about butter making in New York" (311). A model of new farm efficiency, she systematically engages in scholarly self-study; she "went to the library and asked for a list of books on dairy farming. She read with eagerness every one that was given to her, patiently making notes, keeping in her mind the peculiar situation of Old Farm" (245). Soon she is "something rural life had earlier little seen—the expert" (Holt 246–47).

Many Country Life experts believed that the passions of rural individualism desperately needed to be controlled.[22] For example, the Interchurch World Movement (IWM), which was founded and funded by John D. Rockefeller in 1919 and which often worked with the USDA, described the farmer as a " 'sensory child' " suffering from " 'suppressed ideas and desires' " that had to be corrected before they exploded (Neth 106). One social reformer feared that underdeveloped rural life led country boys to " 'while away the long winter evenings talking obscenity, telling filthy stories, recounting sex exploits, encouraging one another in vileness, perhaps indulging in unnatural practices' " (Danbom 32). Urban social scientists thus promoted "mechanisms of order that would control rural society," mechanisms that assumed the existence of industrial farm methods and techniques and that were intended to create in the countryside a middle-class similar to that found in urban areas (Danbom 35).

*Barren Ground* stresses experts' promotion of the control and standardization of farm methods and products, though Dorinda applies this control not only to the farm but also to her emotional life. Bringing the farmer "into intimate relationship with the urban industries which similarly produce their specialties for sale" (MacGarr 162), the New Agriculture gave the farmer "considerable control over the form and substance of the farm product . . . he is shaping his products up to specification, somewhat like the worker in steel, brass, or leather" (Galpin, *Problems* 29). Dorinda standardizes her "creamy butter" for the market by molding it and stamping it with "the name Old Farm beneath the device of a harp-shaped pine"—a nostalgic touch targeting wealthy urban diners, many probably only a generation or two removed from farming (311). Her standardization of Old Farm butter is in lockstep with 1920s advice: "the wise country woman in putting up her products makes them look as nearly machine-made as possible," especially for the "city woman" who "lives in a machine-made world" (Atkeson 117).

Dorinda plans to reclaim Old Farm only after she learns that theory can contain passion. In New York, when the possibility of marriage to Burch arises, she tells Faraday that she is "finished with all that sort of thing," that the idea of sex makes her "sick all over" (237). But soon after, Dorinda struggles to contain her passions at a concert that she and Burch attend. The music reminds her of Old Farm: "rain in the pasture . . . sunsets over broomsedge . . . wild grass [was] burning" (239). During an appassionata, "ecstasy quivered over her, while sound and colour were transformed into rhythms of feeling. Pure sensation held and tortured her. . . . Something that she had thought was dead was coming to life again" (239). In order for her to "stand" this passion, she reduces sensation to analysis, prediction, and explanation by discussing her reactions with Burch (240). Leaving the performance, she tells him that she feels "as if [she] had ploughed a field. It made me savage," a reaction Burch defines as "the pure essence of sensation" (241). The young doctor explains that his own reaction to the appassionata was "little more than an intellectual exercise" because he has "knowledge of the theory of music" (241). Their conversation immediately turns to Old Farm and progressive farming (241–42). Struggling all along to cope with her passions without a method of doing so, Dorinda now has one: the systematization of book farming will create the "cheerful cynicism which formed a protective covering over her mind and heart" (346). Returning fertility to Old Farm, she contains her passion within "the logical barriers of the three dimensions" (366).

Adopting scientific farm methods, Dorinda heals her abandoned fields, both inner and external, both human and geographic. In ways similar to farm reformers who described rotating crops as a way to heal " 'sick' soils" (Anderson 281), *Barren Ground* ties medicine to farming: "Doctor Faraday says [farming] is as much a science as medicine" (242). Even Jason notes that the land "could be brought back to health, if they'd have the sense to treat it as a doctor treats an undernourished human body" (113–14). Nathan Pedlar praises the prosperous James Ellgood as a "first-rate land doctor" who "reclaim[s] some bad land" (275); the Ellgood farm thrives in "a period when a corn-field at Pedlar's Mill was as permanent as a graveyard" (19). Often feeling that her "inner life was merely a hidden field in the landscape, neglected, monotonous, abandoned to solitude, and yet with a smothered fire, like the wild grass, running through it" (12), Dorinda herself becomes a land doctor, successfully reviving Old Farm and Five Oaks, using the advice and investments of urban doctors Faraday and Burch.[23] Trained in a doctor's office and living "under the roof of a great surgeon," she has spent two years in New York learning the "patter of science" (249). Though Frederick McDowell notes that "Dorinda too readily assimilates the knowledge needed to set up a successful farm" (153), her rapid assimilation only underscores "the ease and speed with which certain of these principles can be learned" by all farmers (Davenport 49).

The systematization invoked by scientific agriculture structures Dorinda's analysis, control, and repression of her sexual passions. The narrative itself systematically works to hide "sex emotion" (373); critics point out, for example, that a careless reader can miss Dorinda's pregnancy (Raper 154). Dorinda herself alludes to her miscarriage only obliquely (349). Reworking Old Farm, she sees that the "wild part of her had perished like burned grass. . . . Now, armoured in reason, she was ready to meet life on its own terms" (282). The New Agriculture's dispassionate discipline sublimates her feelings by keeping her "senses . . . benumbed by toil" (312). After acquiring Jason's Five Oaks, "she gave herself completely" to "enriching the land with her abundant vitality"—a task that becomes "a devouring passion" (409). Her efficiency leads her to accomplish what her forebears had not: "They had spent their force for generations in the futile endeavour to uproot [the broomsedge] from the soil, as they had striven to uproot all that was wild and free in the spirit of man" (128). But her work is never done. To be "finished with all that," she must ceaselessly add to Old Farm's reproductive fertility by plowing under her passionate nature (526, 245; see 262).

Industrial agriculture's growth ideal, which glorifies the future by burying the past, serves as a foil for Dorinda's wish to bury her memories of Jason Greylock (Neth 218). For example, walking in a "scarred field" after her father's funeral, Dorinda encounters once again her painful past in the figure of Jason, who complains to her, "You've treated me as if I were the dirt under your feet" (306–7). She does not disagree, telling him that he means "just nothing" to her, and she turns her back on him knowing "the infinite relief of having love over" (308–9). A few pages later, she triumphantly receives her first butter payment, and, echoing many progressive farmers, she declares, "If I didn't live in the future, I couldn't stand things as they are" (312; see 197, 483, 493). The rationality of progressive farming makes her life "one flawless pattern" (347); her celibate "[m]arriage had made, after all, little difference in the orderly precision of her days" (387). And no emotion will halt her business: "Hearts might be broken, men might live or die, but the cows must be milked" (316). After burying Jason "she faced the future without romantic glamour, but she faced it with integrity of vision. The best of life . . . was ahead of her" (525).

*Barren Ground* assumes that the New Agriculture is a positive force; the latest is the best and urban values are a remedy for rural ills. As a text promoting the application of urban industrial techniques to farming, *Barren Ground* is at odds with Jefferson's agrarianism. Dorinda's use of specifically northern sources to create her success also places the novel in opposition to the nostalgic stance of many southern male writers who held fast to an agrarianism that defiantly opposed industrialism, writers whose manifesto would be published in 1930 in *I'll Take My Stand*. The male southern writers who followed Glasgow were not writing in her agrarian footsteps—they were writing to cover her industrial tracks.[24]

*O Pioneers!* and *Barren Ground* warn us not to jump to conclusions about farm women and their relationships to nature. Alexandra and Dorinda do not represent women's mystical unions with the land—at least in any simple, pastoral way. In attitude and practice they are new farmers who happen to be women. Asserting control of the landscapes around them using many methods that Norris's ranchers use, they are bachelor women who have kept themselves, as bonanza owner Annixter might say, free from males. Arguing that these women rework a male-centered pastoral tradition ignores how wedded both are to an industrial agriculture that systematizes and orders the natural world to squeeze from it as many dollars as it can. Understanding that Alexandra and Dorinda are new farmers adds another dimension to the debate over whether or not women treat nature any differently than men. If these novels are any clue, the answer is no.

**And all of them stared after the tractor.—John Steinbeck, *The Grapes of Wrath*, 1939**

CHAPTER THREE

# Disciplining the Farmer

## Class and Agriculture in *The Grapes of Wrath* (1939) and *Of Human Kindness* (1940)

The dispossession of farmers in the name of national economic progress is now an assumption, a natural and inevitable fact of American life (Neth 12–13). Even farmers have learned this lesson well. For example, seventy-five-year-old Mildred Sheppard, a Stone Mountain, Georgia, woman whose dairy farm was "finally swallowed up by suburbia" in August 1994, notes: "The city has moved in on us. . . . It's progress and it's good, but it hasn't been good for us" ("Family's Dairy Farm"). Quoted in a news article, her words immediately follow the assertion that her farm will be "developed," implying that agriculture is archaic land use. The repetition of her words in a boldface inset reaffirms the article's belief that progress is inevitable, though a few, like Sheppard, are inconvenienced.

Appearing in the business section of the (Scranton, Pa.) *Tribune*, the report locates and defines Sheppard's farm with a romanticism targeted at urban readers: "In between strip malls, fast-food restaurants and road construction . . . is a 67-acre pastoral oasis of lush green hills and dense woods barely visible from the fresh asphalt . . . the latest rural throwback to fall victim to suburban sprawl." Suggesting simplicity, bucolic charm, and relaxed living amid a rough-and-tumble cityscape, the description evokes a nostalgia that diverts attention from Sheppard's very real dislocation. Her "lush green hills" belong to a simpler, earlier era for business page readers who assume that strip malls and apartment buildings realize the land's true potential. That Sheppard is the *latest* "victim" confirms that the Sheppard example is both typical and repeating. Farms have gone out of business, do go out of business, and will continue to go out of business to advance urban economic progress. After all, the Sheppard farm is a "rural throwback" that has fallen before superior forces: "Dairy farms like the Sheppard farm have bowed across the country to urbanization, larger farms and growing independent distributors" ("Family's Dairy Farm"). But what makes Sheppard news, for business section readers, is that she has taken so long to succumb, a fact the article's headline boldly declares: "Family's Dairy Farm Finally Swallowed Up by Suburbia."

## The Debate: Yeoman or Entrepreneurial?

In April 1939, three years after Mildred Sheppard started farming at Stone Mountain, John Steinbeck published his epic novel *The Grapes of Wrath*, sparking a storm of controversy that pitted industrial agriculture against migrant farmworkers (Lisca 148–51). A year later, in vigorous defense of industrial farming, Ruth Comfort Mitchell brought out *Of Human Kindness*, a direct reply to Steinbeck's portrayal of migrant exploitation. Whereas Steinbeck's novel is world famous for its exposure of farm labor misery, Mitchell's is a little-known and unapologetic justification of class stratification in farming. Examining the debate between these novels in the context of agricultural history reveals how each attempts to condition readers' responses to class issues within industrial farming.[1]

The Steinbeck/Mitchell debate was no local matter. The novels appeared at the end of a depression decade that saw increased farm mechanization, expanded farm size, and a decline in numbers of farmers (Neth 220). With the nation recovering from a severe economic depression and the world embroiled in war, their debate flared at a crucial moment in American history. By 1940 the nation's agricultural leadership was mulling the challenge of "prevent[ing] the formation of rigid permanent classes" within farming (R. C. Smith 814). What kind of farm community would feed the United States? Would the future farm community be many small family farms distributing wealth horizontally? Or would a few farms whose wealth was divided vertically dominate? Would agriculture finally see a strict class division similar to that of urban industry? The choice between preserving a fluid farm community or expanding a stratified one whose values mimicked urban industrial capitalism fueled sharp controversies within the federal government. In the early 1940s, for example, a major battle waged in Congress over federal Farm Security Administration (FSA) attempts to help the rural poor and private sector American Farm Bureau (AFB) attempts to block such aid. The AFB represented the nation's wealthiest farmers and "had not played any significant part in calling attention to the prevalence of rural poverty" (McConnell 98).

In December 1940 the AFB launched an attack on the FSA, whose task was to lend assistance to small farm families, tenants, and migrants (McConnell 100–101). For instance, the FSA managed migrant camps like the Arvin Sanitary Camp in Kern County, California, the model for Steinbeck's Weedpatch (McConnell 92–93; DeMott, Introduction to *Grapes* xix). The AFB eventually killed the FSA in 1943 by having its appropriations denied, charging, among other things, that the FSA was "communistic" (McConnell 106). FSA administrator C. B. Baldwin understood why his bureau was under attack. Testifying before a congressional committee in the early 1940s, Baldwin stated: "'The choice before the committee is whether the small independent farmer should be given an opportunity to maintain and improve his status or whether these large interests should be permitted to take advantage of the war situation to accumulate large land holdings and to make laborers out of farmers'" (McConnell 113). By congressional decree, the large interests won. Class was now officially an issue on the American farm.

In *Of Human Kindness* Ruth Comfort Mitchell articulates the elitist perspective of California's wealthy growers, the Associated Farmers, one of the American Farm Bureau's "natural auxiliaries of an interested and class character," a group Steinbeck held responsible for much of California's migrant misery (McConnell 125; see McWilliams, *Factories* 231–32; *Their Blood* 10–13; Benson 369). Giving the keynote address at the Associated Farmers' 8 December 1939 state convention, Mitchell delivered "an acid condemnation of Steinbeck's best-selling novel" and called for "drastic action directed at those who would make California a hotbed of poisonous propaganda and worker-employee strife" ("Noted Authoress"; Shillinglaw 148–49). Like the Associated Farmers, Mitchell was " 'anti-Communist' "; in her address she criticized California's administration for advocating "Communist doctrines of collective land ownership" (McWilliams, *Factories* 234; "Noted Authoress" 21). Passages of her novel echo her words at the convention.

One of "California's elite," Mitchell knew migrants through her husband, Sanborn Young (Shillinglaw 148). Owner of a two-thousand-acre dairy ranch near Fresno, Young "welcomed the advent of mechanized farming and defended large-scale California farming techniques and policies" (Barriga 3; see Shillinglaw 148 n. 10). A "notorious strike buster" and avid supporter of the Associated Farmers, Young was a state senator (1926–38) at the height of the migrant exodus to California (Shillinglaw 148; Barriga 3). Mitchell, herself a popular novelist, was "much in demand as a speaker" for Republican political organizations and was "a widely recognized booster for 'The American Way of Life' " who "vigorously defend[ed] the status quo" (Barriga 3; see Shillinglaw 148). In writing her novel, she hoped to "replace one picture of California farm and migrant life with another" (Shillinglaw 149).

The Associated Farmers portrayed agriculture as a business that demanded a disciplined proletarian workforce. In 1926 a California farm spokesperson declared that " 'there is a caste in labor on the farm. . . . We are not husbandmen. We are not farmers. We are producing a product to sell' " (McWilliams, "Farms" 421). Fourteen years later, a U.S. Department of Agriculture economist pointed out that growers "assume that the seasonal workers needed at peak periods must and will be available without any responsibility on their part as to whether there is work enough to go round or what happens to the laborers after the need for them on the farms is ended" (Ham 916). Created in 1933 to pass "anti-picketing regulations," the Associated Farmers worked to "squash agricultural workers' strikes and unionizing attempts before they could seriously upset existing practices" (Kappel 212; see Daniel 251–52). To achieve their ends, the Associated Farmers frequently used violence against strikers (Gregory 157; Kappel 214). Flexing their political muscle in August 1939, they saw that *The Grapes of Wrath* was banned in Kern County, California, libraries (Kappel 212, 214).

Steinbeck first drew the ire of the Associated Farmers in 1936 through a series of articles that damned California's "organized industrial farming" for its treatment of migrant workers (*Their Blood* 2). He distinguished between two farm patterns: "Having been brought up in the prairies where industrialization never penetrated, [the migrants] have jumped with no transition from the old agrarian, self-containing farm where nearly everything used was raised or manufactured, to a system of agriculture so industrialized that the man who plants a crop does not often see, let alone harvest, the fruit of his planting" (*Their Blood* 4). Migrants laboring within industrial farming "are never received into a community nor into the life of a community," even though their work sustained that community (*Their Blood* 2). Immersed in studying the migrants' struggle for several years, Steinbeck felt by 1938 the "urgent need to do something direct in retaliation" to alleviate their suffering (DeMott, Introduction to *Working Days* xxxviii). Though never "fully radicalized," he did put "his pen to the service of a political cause" in writing *The Grapes of Wrath* (DeMott, Introduction to *Working Days* xxxviii).

In *The Grapes of Wrath*, Steinbeck envisions people sharing in an egalitarian human family, rather than isolating themselves as competitive individuals. In one of his most direct addresses to readers, he claims that "the quality of owning freezes you forever into 'I,' and cuts you off forever from the 'we'" (206).[2] Radical social change comes only when the dispossessed realize their selfish natures and unite, when they no longer see themselves as individuals but as part of a larger community—when they share food, clothing, troubles, when they move "from 'I' to 'we'" (206).

In Steinbeck's view, the individual's best hope is to assimilate into a transcendent community that recognizes and helps all equally. For example, Casy, the novel's preacher, tells Tom Joad: "Maybe all men got one big soul ever'body's a part of" (33). Peter Lisca points out that as the Joad "family unit is breaking up, the fragments are going to make up a larger group. The sense of a communal unit grows steadily through the narrative" until at the end it encompasses all people (172). Tom Joad, for instance, moves from self-absorbed and "individualistic" to asserting, as Lisca observes, "his spiritual unity with all men" (173). More poignantly, the novel's closing tableau leaves readers with Rose of Sharon sharing her life-giving milk with a starving stranger (619).

But the novel's end ultimately serves industrial agriculture. Though for many the end offers hope—the Joads see that they share with all a common humanity—Rose of Sharon's enigmatic smile still leaves them with no food, no job prospects, and little hope for physical survival. At best, they will continue traveling from ranch to ranch, following the crops, though now accepting more easily, because of their new understanding, their plight and position within California farming. Owners still own them.

Taking refuge in the Joads' realization, the middle-class Steinbeck could envision no radical change in the class structure of California agriculture. For example, reporting in 1936 on migrants' poor conditions, he warned the state that it is careening toward revolution by "gradually building a human structure which will certainly change the state, and may . . . destroy the present system of agricultural economics" (*Their Blood* 5). To avoid migrant "fury," he argues, keep the system, but make workers amenable to it by creating farm labor unions (*Their Blood* 33). This is exactly what happens in *The Grapes of Wrath*. But with the Joads safe in their newfound community, industrial agriculture emerges from the novel triumphant, leading readers to understand that the Joads' "progress" from farmers to sharecroppers to migrants is inevitable.

Mitchell's story is told mainly from the perspective of "City slicker" Mary Ashley Banner, the daughter of an aristocratic San Francisco family who must learn what it is like to live on a modern, middle-class American farm (14).[3] The Banners are well-established citizens: "San Joaquin Valley pioneers, third generation in California; plain people, poor people, proud people; salt of the earth" (5). A family of true patriots—daughter Sally's imaginative nickname is "Star-spangled"—their surname "has a sort of—of a *marching* sound!" (64, 259). Family members, "proud only of their energy and thrift and robust Americanism," hold dearly to "accepted standards of public and private conduct, to sane thinking, to a sense of clan loyalty" (5, 233). Though they earn a living from commercial dairy ranching, by novel's end they have moved on to discover politics and oil (355).

The novel's plot is set in motion when Sally Banner runs away from boarding school to marry Lute Willow, a guitar-playing Okie hired hand, whose main traits are, according to Sally's father, Ed, "his dumbness, his shiftlessness" (67). Under the guiding hand of his new wife and father-in-law, Lute eventually learns to be an Associated Farmer. Meanwhile, Sally's brother, Ashley, falls under the spell of "reds," first his history teacher, then a voluptuous labor agitator—only to be sorely disillusioned. Though Ed Banner runs his son off the farm, Ashley learns his lesson well: he works hard, returns—hating communists with a passion—and nominates his father for state senator. Already trained in "special work along scientific dairy lines," the restored Ashley will command the farm while his father helps rule the state (214).

Answering Steinbeck's vision of an essentially classless, cooperative human family, *Of Human Kindness* argues that migrants must be made obedient citizens of a class-stratified entrepreneurial farm community. Defining American farmers as individual competitors whose basic unit is the nuclear family, Mitchell claims that Dust Bowl refugees need discipline; they must be taught how to behave in California's industrial farming world. Carey McWilliams, an early reviewer of the novel, alludes to this discipline when he sarcastically points out that in her "partisan document" Mitchell's Banners are "a yeoman farm family of the San Joaquin Valley," who are "sociological saints," who "use force, yes, but it is benevolently applied" in the service of the Associated Farmers' understanding of agriculture ("Glory").[4]

Anthropology offers precedent for dividing farms into yeoman and entrepreneurial types. I adopt these terms from Sonya Salamon's study of midwestern farm communities, *Prairie Patrimony* (1992). Salamon argues that within midwestern farming there exist "two cultural systems" (3): "the German cultural pattern *yeoman* and the Yankee pattern *entrepreneur*" (7). She concludes that these cultural patterns shape the wider community: "For yeomen, the practice that defines the relationship of family to community is a commitment to continuity. . . . For entrepreneurs, the practice that defines the relationship of the family to the community is a commitment to maximizing individual financial returns" (229).

As I define it here, the yeoman pattern is rooted in mutual dependence and assistance among people who recognize shared responsibilities. People in this community value personal relationships in preference to economic or political ones; relations among them move in horizontal directions (Neth 41). The entrepreneurial pattern prefers vertical economic and political relationships to personal ones: farming is a business and only the most efficient farms and families should survive. The yeoman community is rooted in neighboring, a "general equality of exchange. Neighboring placed economic exchanges within social relationships that often overshadowed their economic nature" (Neth 41). Unlike the yeoman neighborhood, the entrepreneurial community need not consist of nearby farms—in fact, the large size of most entrepreneurial farms precludes neighborliness. Whereas the yeoman community is a mosaic of interdependent small family farms, the entrepreneurial is a pyramid of urban-dependent agribusinesses.

## "I Own a Farmall"

Although most of today's farms—whether yeoman or entrepreneurial—have been thoroughly disciplined to accept industrial agriculture, such industrialization is not complete. For example, many Amish stubbornly refuse to adopt electricity and modern machinery, yet they continue to harvest comparable crops per acre using horse-drawn equipment. In their refusal, they see as one the needs of farm and community: "the Amish think of 'the community as a whole'. . . . The wholeness or health of the community is their standard. And by this standard they have been required to limit their technology" (Berry, *Unsettling* 212). To keep themselves whole, they choose to restrain themselves, something most American farmers seem incapable of doing when faced with a new technology. Perhaps the most significant choice that the Amish made to limit the impact of technology on their community was their 1923 decision to ban tractors from fieldwork (Kraybill 172). While the Amish worked with horses in the 1930s to keep farming, other farmers rode tractors to their ruin.

Nothing in the 1930s symbolized farm industrialization and class stratification more than the all-purpose tractor, the Farmall, first marketed by International Harvester Company in 1925. A major breakthrough in farm technology, the Farmall's tricycle design allowed it to cultivate row crops in addition to plowing and hauling equipment. Even as thousands of farmers went bankrupt in the 1930s, the Farmall "sold so well that it eventually changed the image of the tractor" (R. C. Williams 88). According to the U.S. Department of Agriculture 1940 *Yearbook*, "Three-fourths of all tractors sold in the United States in 1937 were general-purpose tractors" (Kifer, Hurt, and Thornburgh 513). The Farmall was useful to all farmers, regardless of the farm they owned, the type of soil they worked, or the type of crop they planted—the only thing they needed was money to buy it. More than anything, the tractor brought home to vast numbers of farmers a new type of farming: industrial.[5]

Because of tractor power, farmers had to change how they farmed (McMillen 9). Though many kept a team of workhorses for a season or two to complement their tractor or to see how the machine would work out, most farmers soon sold their horses and relied on mechanical horsepower (R. C. Williams 88). Tractor advertisements encouraged this as late as 1939 (see Fetherston 31). With reduced labor costs—"By 1935, the tractor was saving some 165 million man-hours per year in the United States in field operations alone"—tractor farmers could afford to plant more crops, which in turn created surpluses that resulted in lower prices (R. C. Williams 153). Those who could not afford a tractor were soon caught between high labor costs and low prices, forcing them into tenancy or out of farming altogether. Hardest hit of all were hired laborers and sharecroppers. Because the tractor reduced labor costs so dramatically and so quickly, some blamed it for the Great Depression (R. C. Williams 155, 153).

For thousands of farmers a new tractor symbolized a rise in status—owners were modern, progressive, sophisticated. But for thousands of others it was emblematic of their physical and psychic dislocation. For example, hemmed between economic depression and nature's wind and drought, many Plains farmers castigated the machine for "tractoring" them from their homes. The starkest image of this might be Muley Graves's flashback in John Ford's 1940 film *The Grapes of Wrath*, when moviegoers witness a Caterpillar demolish the Graves home. Some of the dispossessed were former farm owners and sharecroppers who ended up as migrants, like Steinbeck's Joads, following harvests season after season, members of no fixed community.[6]

As more and more farms industrialized, fluid class relations within farm communities hardened. The agricultural ladder of the nineteenth century—the idea that one could step from hired hand to tenant to owner—virtually disappeared by 1940 because farming had become increasingly a capital-intensive venture, primarily due to tractor power. The U.S. Department of Agriculture (USDA) actively equated independent farming with tractor ownership. For example, a 1940 USDA drawing of the agricultural ladder depicts a sharecropper following a mule and an owner-operator driving a tractor (R. C. Smith 812). By 1940 the USDA reported 1.6 million tractors in use on American farms, almost double those in 1930 (Kifer 513). Amid ever-widening farm consolidation and rising numbers of tenants, depression-era farmers defined themselves progressively as individuals competing in national and global marketplaces, rather than as neighbors living in a single locality (Neth 219–20; Maris 888–89). With the pace of tractor sales crushing the hopes of those who could not afford them, tractors widened the gap between the well capitalized and the cash-strapped, remaking rungs on the agricultural ladder into class ranks. And what sold tractors was an advertisement that pitched entrepreneurial values to the nation's farmers.

In the same month that *The Grapes of Wrath* appeared, International Harvester (IH) advertised its Farmall model 14 and model 20 in the opening pages of the *Country Gentleman*, a widely circulated farm journal. The advertisement teaches the values of industrial entrepreneurial farming by defining a constituency at odds with an interdependent farm community: the ad's target farmer is individualistic, an owner interested in saving time and labor, a money-conscious man attracted to technological power. Bannering the ad is a large quote from a Farmall owner—one of the many "enthusiastic owners who have proved the value of their Farmalls": " 'THANK YOU for Building a Great Tractor—the FARMALL.' " A red arrow encasing the quote—red is IH's main company color—points to a male farmer, presumably the owner, leisurely looking left at the ad's text as he sits astride a Farmall 20, whose front end extends toward the viewer/reader. The ad declares that International Harvester has received "many letters" from happy owners: "thousands have taken the trouble to write." The assumption here is that farmers with Farmalls have saved enough time to bother writing, and a shorter workday is the hope of many tractor buyers (R. C. Williams 132). Linking machine and time again, the ad claims that "each added feature or improvement has brought new praise"—presumably, more letters.

The ad also targets a farmer who values speed: "*You can do* A REAL JOB OF CULTIVATING *with the Farmall 14 and quick-attachable cultivator.*" Saving time is an essential feature of an advertisement aimed at the businessman who knows that time is money. The first of the Farmall's selling points in a numbered series of ten is that the Farmall has a "patented automatic steering-wheel cultivator gang shift. Clean cross cultivation at 4 miles an hour." Pictured in inserts are two farmers working alone in their fields, one planting, the other cultivating, both with Farmall 14s. Suggesting the tractor's potential for swelling farm size, one man works a field stretching to a limitless horizon, reminiscent of the bonanza farms of chapter 1.

Though all tractors originally were gray, by 1940 most were "brightly finished" in various colors that "presented an attractive sight in fields" (R. C. Williams 99 n. 48). International Harvester introduced Harvester red with the Farmall 20 (R. C. Williams 97 n. 12). To capture the farmer's gaze, IH promoted Farmalls as "handsome red tractors," following the lead of Allis-Chalmers in giving tractors " 'sex appeal' " by painting them "eye-catching

Farmall tractor advertisement. From *Country Gentleman*, April 1939.

and identifiable" colors (R. C. Williams 95). IH advertising coupled this sex appeal with power. For example, in the January 1939 issues of *Country Gentleman* and *Hoard's Dairyman*, an IH ad notes, "If you want *beauty*, insist on the *useful* beauty of FARMALL power and performance . . . insist on the RED TRACTOR." The message is clear: the most potent farmer rides the reddest tractor. Be productive. Buy an IH.

The Farmall was such an asset to progressive farmers that "the farmer's

Farmall tractor advertisement. From *Country Gentleman*, January 1939.

proudest boast is 'I Own a Farmall.'" The capitalization of "own" was no accident; ownership distinguished the most powerful class of farmers from those behind the times, that is, those still following horses. Using peer pressure to get farmers to buy, the ad created a nameless, numberless group of Farmall owners: "these owners will tell you it pays to pick the *genuine* McCormick-Deering Farmall." Farmers will want to buy a Farmall, the April ad suggests, to keep up with farmers "all over the land [who] are at work with their Farmalls, enjoying the power and performance" of their machines. Such farmers were not planting or growing crops; they were "building their crops with Farmalls." Building suggests manufacture, not biological process, likening agricultural practice to the industrial systems of Chicago-based International Harvester.

## Tractored Out

A symbol of status and progress for some, the tractor was a sign of dispossession and defeat for others. In Steinbeck's novel, migrants have been forced from Oklahoma because they have either been "dusted out" or "tractored out"—either nature or technology has forced them to the road (12). But the novel does not argue that technology itself is the enemy—it is the use of that technology: "Is a tractor bad? Is the power that turns the long furrows wrong? If this tractor were ours it would be good—not mine, but ours. If our tractor turned the long furrows of our land, it would be good. Not my land, but ours. We could love that tractor then as we have loved this land when it was ours" (205). In other words, if the *community* had chosen to use the tractor, the machine would have been welcome; approval suggests the possibility of retraction if the machine threatens community survival.

Tractor and community were not necessarily at odds. In fact, the federal government attempted to make tractors "ours" rather than "mine." In 1938 the Farm Security Administration supported a cooperatively manufactured tractor in the "Federal subsistence community" of Arthurdale, West Virginia ("Tractor 'Co-ops'" 33). According to a *Newsweek* report of 5 September 1938, "a farm cooperative organization has contracted with an association of homesteaders to assemble [the] tractors" ("Tractor 'Co-ops'" 33). The project would have created a tractor founded in community, but it was denounced by "Republican congressmen . . . as an attempt to 'Sovietize American industry'" (R. C. Williams 175). Though the effort eventually folded, "the project offered the prospect of real competition in the industry" (R. C. Williams 175). Ironically, at about the same time International Harvester, then the nation's largest tractor maker, was rebuked by the Federal Trade Commission for its "structures" and "practices" in "a searching monopoly investigation" (R. C. Williams 175; "Tractor 'Co-ops'" 33). Cooperative ventures such as the FSA's plan had no chance in the face of IH's national network of production, distribution, and advertisement.

In interchapter five of *The Grapes of Wrath*, a tractored-out tenant learns all too well the power of the entrepreneurial farm community. In this chapter Joe Davis's boy and the foreclosed tenant argue over Davis's decision to become a tractor driver, essentially an argument between neighboring and individualism, between yeoman and entrepreneurial communities. Taking a break from "raping methodically, raping without passion," Davis eats lunch without sharing anything with the tenant's hungry children (49). There only to harrow the land, plant a crop, take his pay and move on, Davis works like an automaton; he "did not look like a man" (48). On the clock and with orders to keep the lines straight, he eats at noon and is "going through the dooryard after dinner" to topple the tenant's home (51). In the journal he kept while writing the novel, Steinbeck worried about getting this chapter just right: "Got to because this one's tone is very important—this is the eviction sound and the tonal reason for the movement" (*Working Days* 23).

Sitting high in his seat while disputing the tenant, Davis looks down on the dispossessed, arguing for values characteristic of IH ads. As a tractor driver, Davis has elevated himself: "I got damn sick of creeping for my dinner—and not getting it"; along with his new perspective he receives "three dollars a day, and it comes every day" (50). Refusing to share his midday meal, Davis points out that "times are changed"—that is, the tenant must keep up with progress or suffer the consequences (51). And this tractor driver has learned his place; to question progress is to be labeled dangerous: "Big shots won't give you three dollars a day if you worry about anything but your three dollars" (51). Assuming that a family is an isolated economic unit, Davis rejects the tenant's argument for a neighborhood of shared burdens and responsibilities; he tells the tenant that opposing his own displacement will not "feed the kids. . . . You got no call to worry about anybody's kids but your own" (51). Finished eating, Davis shows the power of his argument by crushing the tenant's home "like a bug" (52). A disciplined worker, he has his orders; he has "got to keep the lines straight" or lose his job (51).

In response, the tenant defends community by arguing for neighborhood: "But for your three dollars a day fifteen or twenty families can't eat at all. Nearly a hundred people have to go out and wander on the roads for your three dollars a day. Is that right?" (50). The tenant does not dispute that times change; he simply challenges the kind of change by arguing that entrepreneurial agriculture so upsets the necessary balance between humans and nature that it destroys one's humanity. If a man has "property he doesn't see . . . or can't be there to walk on it—why, then the property is the man" (50–51). Dismissing the argument, "The driver munched the branded pie and threw the crust away" without a thought for the children he beggars (51).

Whereas the tenant places community first, Davis's entrepreneurial perspective divides people. For example, Connie, Rose of Sharon's husband, believes that learning industrial farming will realize the couple's dreams of a home. But his material hopes so rule him that he cuts the less tangible ties of family responsibility. Seeking to "get on [his] feet," he runs away, abandoning his pregnant wife, telling her it "would a been better maybe to stay home an' study 'bout tractors. Three dollars a day they get, an' pick up extra money, too" (343–44). His words repeat Davis's and implicate him in the destruction of the community he married into and divorces himself from. Already thinking little of his wife and child, he will soon learn what Davis already practices: to gain a home is to knock another's down.

Davis's attention to keeping lines straight was no aesthetic consideration; saving time and money was involved. With the introduction of the tractor, fields were often combined and redesigned with right angles to allow more room for turning (R. C. Williams 177). The straighter and longer the lines, the more time the farmer saved planting, the more money he reaped in reduced labor costs. But the resulting farm geometry left no margins. Hedgerows and irregularly shaped corners were eliminated and brought under cultivation, leaving no room for small animals and birds (R. C. Williams 177). The new farm landscape was no place for human emotion either; the land was "not loved or hated, it had no prayers or curses" (49). To underscore how the new configuration killed the connection between nature and the human soul, Steinbeck emphasizes Davis's rulerlike work: "straight across country [the tractor] went, cutting through a dozen farms and straight back" (48); the driver's threat to destroy the tenant's home: "Got to keep the lines straight" (51); his cold-blooded attitude in doing so: "The tractor cut a straight line on" (53).

In *Of Human Kindness* the tractor is conspicuous for its near absence. Disputing claims that Okies were driven off their land, Associated Farmers defend tractors: "These migrants weren't all dusted out nor tractored out" (292). The novel's token Okie, Lute Willow, after all, did not go to California because he was "dusted or tractored out"; he went west because "California was just a'burstin' and a'boilin' with fat an' easy jobs!" (117). Tractors also *rescue* the trouble-stricken. One Banner hired hand tells Mary that if

Sally "don't like that boardin'-school a bunch of us'll drive the tractor up there and get her!" (99). In a story that responds to the end of *The Grapes of Wrath*, a migrant camp nurse tells how a pregnant worker was carried from a flooded cabin aboard a tractor (333–34). In the nurse's story, the tractor is firmly controlled by the men who use it and not the other way around, as Steinbeck pictures the man-tractor relationship: "The ranch foreman ran it . . . and a fellow sloshed ahead of us with a lantern" (333–44; see *Grapes* 48).[7]

Like the absentee owners in *The Grapes of Wrath* who want to stretch their bottom lines, the entrepreneurial Banners "hunger" to straighten their property lines, and they do so by investing in more land (123). Though Mitchell describes them as simple yeomen, the Banners are not as poor as their neighbors believe—though the neighbors' perception may be skewed by their own wealth (5, 135). A certain " 'land look' . . . was always there when they were discussing the wisdom of buying forty acres . . . to straighten out the line on the west edge" (123). The history of their ranch is one of steady expansion from fifteen acres to eight hundred: "We've tacked on pieces here and there to straighten our lines, to get more pasture as the dairy business expanded, and to-day we have close to eight hundred, clear, alfalfa, oats, corn for the silos, permanent pastures" (291). Describing their purchase of Morton's Forty as an aesthetic move, the Banners redraw the landscape with the stroke of a pen (135). Studying his farm from an upstairs window in his new house, Ed Banner exclaims: "Golly, what a difference a few feet up in the air make! . . . Sort of get the whole pattern from here . . . we've never had a look at it like this in all the years" (132). Ed's obsession to expand and order his holdings so characterizes him that Mitchell likens him to his fields: "straight, square . . . looked up to as one of the community's best citizens" (135); he is "a straight shooter" (153). A spokesman for the Associated Farmers, he is also making a beeline to the state legislature.

## The Long Perspective: Education and the Vertical Community

Mitchell returns again and again to the Associated Farmers' expansive perspective to teach readers that Steinbeck's view of the migrants' trouble is myopic and insufficient (133). As Mary listens to Ed marvel at his enlarged farm, she envisions the new house as "a symbol—the Banners getting above their stubborn acres, above their problems, getting the long look, the perspective" (133). The message to the reader: those who accept Steinbeck's view of the migrants' trouble do not have the proper perspective on events; they must consider another point of view, the growers', the Associated Farmers'. Public relations and good advertising—Mitchell's public speeches and her novel, for example—will tell the right story. The novel confirms this: since the urban East has "a pretty dark picture of California farmers . . . [as] 'Bankers with pitch forks,'" banker Burton Doane calls in writer Kent Dexter to set the record straight (258). Adopting the growers' interpretation of events, *Of Human Kindness* "dovetail[s] with the Associated Farmers' own campaign" against *The Grapes of Wrath* (Shillinglaw 150). Readers who have looked up from the bottom with the Joads can now look down from above with the Banners.

Such long looks teach readers to rank problems by their relative importance: migrant misery pales in comparison to the necessity of providing food for an ever-urbanizing nation. And attention to sizable perspectives abounds in the novel. Wealthy grower and Associated Farmer leader D. K. Starrett, a man who "stepped on a plane for New York as casually as the rest of [the growers] climbed into their dusty pickups," owns "the famous Valley Vista with offices in San Francisco and Los Angeles" (185). San Francisco banker Burton Doane advises Mary "to stand off and get the perspective, the long view" to understand the migrant problem (220–21). Associated Farmer Humphrey Pastor tells Kent Dexter, whom he calls "Mr. Easterner": "You've got to get under the surface, you've got to take the long look. A hundred factors . . . have combined through the centuries to bring these people—these exiles, these radicals—to our fields" (295). Mary gains the most extensive perspective on her personal troubles while flying to San Francisco: "The closeness to infinity put finite things back into scale; looking down on the earth lights and up at the stars she knew comfort, reassurance: fears fell away into the darkness" (310). Apparently, the fewer details you see, the better off you are, an attitude reminiscent of Norris's bonanza ranchers.

Elevated awareness can solve all problems. For example, at a crucial moment in the novel Mary's raised perspective breaks up a labor strike and saves an ambulance driver's life. From an upstairs window in the "main office" of Starrett's ranch, she looks down on striking migrants who confront the police after refusing to leave the ranch or work its fields (230). As if calling attention to the significance of her proper perspective in this scene, Mary wonders about her son's clouded vision after falling under the spell of communist agitators: "The boy had lost his sense of proportion and of perspective for the moment" (233). When strikers bar ambulance attendants from taking away a pregnant migrant, a commotion breaks out. But the crowd soon relents and the ambulance speeds off, only to strike and kill leftist history teacher Pinky Emory when she throws herself in the ambulance's path, sacrificing herself for the communist cause. When Emory is killed, bedlam breaks out among the strikers, who are convinced by labor organizers that Starrett ordered her death. Accused of murder, the ambulance driver cries out for someone who shares his perspective: "didn't anybody see it happen?" (242). Of those who do see Emory die, Mary is the only one who really understands what happened because she has seen and heard everything from her lofty point of view.

The only one with the proper perspective—which, not incidentally, originates from the nerve center of Starrett's ranch—Mary answers the driver's plea for help: "I saw it happen! . . . I could see everything" (242). The migrants below simply do not have the best view; as Mary comes forward, there is "a jostling and craning of necks to see who she was" (242). In her most heroic moment, Mary destroys the strike when she accuses Ashley's seducer, the communist rabble-rouser Black Widow, of murdering Emory: "I know what made it happen! *You* did!' . . . This woman is a cheat and a liar! She didn't love this poor dead creature! She scorned and abused her; I heard her!" (243). Mary saves the day for the Associated Farmers because the crowd was, "for the moment at least, deflected" (243). Soon police reinforcements arrive, and Starrett's men clear the ranch for workers who will quietly harvest his crops (244).

In contrast, long perspectives in *The Grapes of Wrath* deceive. At the novel's center, the Joads enjoy a misleadingly pastoral perspective of the San Joaquin Valley from high in the Tehachapi Mountains, their first good look at California farming: "The vineyards, the orchards, the great flat valley, green and beautiful, the trees set in rows, and the farm houses" (309–10). What has created this geometric order is not a neighborhood of small farms, as they may assume, but a hydraulic, industrial agriculture hoping to hide violence, class hatred, and starvation wages. Though the great valley promises much from this height, once the Joads are down in the valley, among workers, and in the fields, their perspective changes from hope to horror. They soon learn their lowly place in California's entrepreneurial farm "community."

The Joads' initial confusion in confronting California farming is represented in their reaction to the Hooverville, where they first seek shelter: "There was no order in the camp; little gray tents, shacks, cars were scattered about at random" (328). Floyd Knowles, who knows California farming firsthand, teaches Tom Joad about working in the industrial valley. Work is found on the move; wages are low because migrants compete fiercely for jobs; work ends quickly when a crop is picked; migrants are expected to move on when the job is done; you can be tossed in jail and blacklisted for leading a strike (334–36). Floyd tells Tom: "I know ya jus' got here. They's stuff ya got to learn. If you'd let me tell ya, it'd save ya somepin. If ya don' let me tell ya, then ya got to learn the hard way. You ain't gonna settle down 'cause they ain't no work to settle ya. An' your belly ain't gonna let ya settle down. Now—that's straight" (355).

Floyd's greatest lesson is that grower violence disciplines workers into silence and acceptance. To avoid the damage inflicted on the Hooverville's "mayor," Floyd tells Tom that he must feign ignorance: "when the cops come in, an' they come in all a time, that's how you want ta be. Dumb—don't know nothin'. Don't understan' nothin'. . . . Be bull-simple" (338). To speak out is to be struck down. The supposed migrant ignorance that *Of Human Kindness* hopes to check with "GOOD OLD AMERICANISM" is an intelligent reaction to life-threatening situations such as the shootings, beatings, and burning visited on the Hooverville that night (147; see *Their Blood* 12–13).

To survive, Steinbeck's migrants rely on a horizontal community that values social relations over economic ones. After purchasing overpriced food at the strictly entrepreneurial Hooper Ranch, Ma notes the good deed of "the little man" behind the counter who buys her some sugar: "I'm learnin' one thing good. . . . If you're in trouble or hurt or need—go to poor people" (513–14). A direct contrast to the Hooper Ranch, Weedpatch, the government camp where the Joads seek refuge, is the ideal. In addition to sharing responsibility for running the camp—the Central and Ladies Committees, for instance—residents share food; Tom eats breakfast with a family he has never met before: "Had your breakfast? . . . Well, no. . . . Well, set down with us, then. We got plenty—thank God!" (396). The men of this family even share scarce work with him (397).

Learning to share begins with the young at Weedpatch. Ruthie Joad learns a painful lesson in social relations when she selfishly tries to enter a croquet game without so much as asking to be admitted. Though the others agree to let her in the next game, Ruthie won't wait, snatches a mallet, and demands to play. The others then let her have her way, but she plays alone: "The children laid their mallets on the ground and trooped silently off the court" (434). Ruthie tries playing by herself, but finally runs home to cry after realizing that the others will not share in the game because of her selfish attitude. But the children do include Ruthie when she returns: "The watching lady warned them: 'When she comes back an' wants to be decent, you let her. You was mean yourself, Amy' " (434). Steinbeck here envisions a community that values human relations above and beyond Ruthie's—and Herbert Hoover's—rugged individualism.

As an answer to Steinbeck's hope for a horizontal community, Mitchell's novel defines agriculture as a community of strong, independent individuals who discipline themselves and others in vertical relationships based on monetary exchanges. For instance, Mary's extended flashback about her initial arrival at the Banner ranch is framed by the remarks of a hired hand, "an elderly neighbor . . . who had lost his own ranch" (7, 56). From the Banners' perspective, well-fed hired hands like this one are "good-natured and cooperative. Square treatment paid good dividends" (268). There are other examples: doing Mary a favor, Burton Doane uses his financial connections in the valley to find Okie Lute Willow a steady job (118). A woman who nurses Mary's Cousin Isabelle is the "paying guest" (301, 305, 310). To get trespassing migrants to move off Banner land, Mary bribes them with a ten-dollar bill (346). The Banner farm's progress is partly underwritten by an inheritance Mary receives that gives her a monthly allowance of $100; when the independent Ed finds out, he exclaims, "Gee, I guess I did marry you for your money!" (47).

## Disciplining the Farmer

To assimilate properly into California's entrepreneurial farm community, displaced Plains farmers like the Joads need to be reeducated. In *Of Human Kindness* migrants are ignorant lowlifes: Mary tells Ashley that Okies are no good "because they're a certain kind of people—unfortunate, yes, and not to blame for their misfortunes, of course, but ignorant and shiftless" (81). Later, she tells friends: "They're here now [the migrants], and isn't it sensible to teach them and feed them properly and put them in shape to be healthy, intelligent citizens?" (150). Humphrey Pastor, president of the Associated Farmers, concurs, telling journalist Kent Dexter that all anyone needs to do is "just give them time and a fair chance—our health service, our schooling—they make good citizens" (293). Ed Banner agrees, pointing to his son-in-law as an example of a transplanted Oklahoman who is "learning to be a good farmer. And he's going to be an Associated Farmer" (293). Pastor adds, "That's the history of lots of them. The best of them, the ones we can assimilate" (293). As Mary points out, proper discipline will "make citizens *for us*"—that is, Associated Farmers (150, emphasis added).

Steinbeck did not disagree; he understood industrial agriculture as an inevitability, not as a choice. In 1936 he claimed that two alternatives existed: migrants "can be citizens of the highest type, or they can be an army driven by suffering and hatred to take what they need" (30). To avert the latter, he urged a change in the growers' perspective: if "future farm workers are to be white and American . . . a rearrangement of the attitude toward and treatment of migrant labor must be achieved" (*Their Blood* 28). He does not question the necessity of an agriculture based on migrant labor; he only points out that to run smoothly such an agriculture must be kinder, gentler. Suggesting his acceptance of industrial farming, he defines Dust Bowl migrants as "anachronistic" because they have moved from self-sufficient farm communities to the capital-intensive farms of California. Migrants have come to a "new" life, completely different from their old lives in "little farm districts where democracy was not only possible but inevitable" (*Their Blood* 4). He warns that every effort must be made to settle "this new kind of migrant" because it is not in the migrant's nature to wander; within each "is one urge and one overwhelming need, to acquire a little land again, and to settle on it and stop their wandering" (*Their Blood* 2–3). In effect, Steinbeck claims that if they must be farmworkers, they can at least be homeowners. Can't they? But California growers wanted workers, not neighbors: Okie land ownership of any kind was not their first suggestion to solve the migrant crisis—as "one large speculative farmer has said . . . the success of California agriculture requires that we create and maintain a peon class" (*Their Blood* 3).

To function at all, the agriculture in both novels demands an underclass that owns no land and can be part of no community—a problem that *Of Human Kindness* avoids and *The Grapes of Wrath* cannot solve. At the "bottom of the economic scale in farm work was the migrant laborer," people like the Joads, who by definition were unassimilable into a fixed community because they were constantly on the move, with no fixed abode (R. C. Williams 156). Ironically, migrants were "a prerequisite of mechanization and often a casualty of it" (R. C. Williams 156). Tractor historian Robert C. Williams points out:

> The migrant farm worker was necessary to fill gaps that resulted from the uneven pace of mechanization. The tractor, for example, minimized the amount of labor needed for most aspects of cotton production, especially plowing, planting, and cultivating. But the tractor did little to reduce the need for hoe hands or cotton pickers. So in hoeing season and picking season, there was at least as much demand for hand labor as in the days before tractorization. Incomplete mechanization had thus created a seasonal imbalance in labor requirements, and the migrant filled the need. (156)

For the large growers populating *Of Human Kindness*, migrant assimilation means molding citizens who will quietly learn to accept California's seasonal farm labor system; good peon behavior makes for good state citizens. For example, after describing how his "good steady bunch of folks" turned away "agitators" because they realized they were "getting fair pay and fair treatment," the owner of the ranch where Mary discovers Ashley describes his migrants as "nice, decent people, doing better'n they did at home, they tell me. I believe they'll make good citizens if we give 'em a chance" (335–36). Sally and Lute work cotton and fruit fields for a grower who is "kindly, indulgent," whose wife "keep[s] an eye" on them because "they're nice kids; just the kind we like for steady help" (121). Obviously, the young newlyweds have learned their place, at least as far as these owners are concerned: "they behaved themselves and did their work. . . . If they hadn't, they'd have been thumbing rides somewhere" (197). A longtime resident of the valley affirms the importance of good migrant behavior: "When I was a young-one ranch-hands were pretty meek and mild, and nice folks," the implication being that today's hands are not so well mannered (149). If migrants would just learn to work hard, ask for little, and keep quiet (bull-simple), their lot will improve. Defining this state of affairs as natural, growers make migrant behavior the root of the problem, not the work conditions or the farm system workers labor within.

Like International Harvester and the American Farm Bureau, *Of Human Kindness* marries urban and rural interests through proper education.[8] Land grant university ag education unites upper-class urban Mary with middle-class rural Ed. Like Annixter in Norris's *The Octopus*, Ed is a well-trained farmer; at the Ag College at the University of California, Davis, he was a protégé of the chair of the Animal Husbandry Department. The chair, who introduces Mary to Ed, regards Ed as "a pretty good specimen" for her (11, 14). Married within ten days, Ed takes his "City slicker" to his mother's ranch, determined to make a rancher of her, which is no easy task (14). When Mary arrives, she is "so soft and helpless, so ignorant of ranch life. I'd never been taught anything useful" (253). Learning as much as possible to be "a good ranch wife," she succeeds in becoming "utterly molded and blended into her environment" (253, 251). Mary learns the New Agriculture so well that its values are reflected in her language. For instance, to convince Ed's mother, Sarah, that she is still valuable to the farm following her stroke, "[Mary] had decided that she would share everything possible with her, make her feel she was still a working member of the family firm, still on the Board of Managers; still needed, necessary" (216). By late in the novel, Mary cannot dissociate herself even linguistically from the biggest ranchers: "There mustn't be any violence. For [migrants'] sakes, first, of course, and for our sakes—for the Associated Farmers, I mean" (342).

Her assimilation into the entrepreneurial farm community does not destroy her urbanity, however, nor is she a drudge like the battered Edna Cosgrove. Mary has her own car, a "spry coupé . . . her 'Magic Carpet,'" which she drives to San Francisco (87). She keeps "a foothold in her old world" because "Burton passed his magazines on to her—literary reviews, journals of opinion, digests—and the best of the new books, fiction, travel, biographies, philosophies, and challenges, and demanded comments which attested thorough reading" (78). Training her to maintain her wide perspective, banker Burton mails her his version of a "college education" (78). In retaining her upper-class refinement, Mary mainly learns to be an urban middle-class farmwife. She is the urban agrarian dream come true.

In contrast to the straitened Joads' realization of a larger human family, Mary Banner believes in a tight "Family," so much so that she wants to herd hers "into a stockade of security and stand guard over them" (70). Unlike the Joads, the Banners are a small family—husband, wife, daughter, son, along with a grandmother who soon dies. The Banner ranch needs no large family because Ed adopts "modern machinery, modern improvements, modern methods"—all things that save labor—things readers never see in action (172, 28). Unlike the Joads, the nuclear Banner family ends the novel intact and at home. Whereas the Joads rely on an extended family to stay alive, the Banners rely on farm expansion and technological innovation to feel secure (24, 85, 127, 170). And unlike the Joads, when the Banner children leave home, they know that they have a home to return to.

To rejoin the entrepreneurial farm community and inherit the Banner farm, left-leaning Ashley must be reeducated. Forgoing an opportunity to get the right education in San Francisco, he falls for the wrong kind when he acts upon the ideas of left-wing history teacher Pinky Emory (187). But hard work—and seeing his communist seducer (the Black Widow) in bed with another activist—soon cures him of his leftist politics, his "bewitchment" (338). He returns to the Banner ranch after laboring on another farm, where readers see him "working so fast" (336). Like his sister, Sally, he finds employment with a kind grower, who notes that his "wife took a fancy to him right off, has him eat with us" (336). (Even Ashley's parents do not eat with hired hands [96].) That Ashley's experience in the field of hard knocks has been edifying becomes clear when his employer exclaims—"and spaced the words for extra emphasis. '*Does—he—hate—Communists!*'" (336). His reeducation straightens him out and returns him to the family fold a changed man: "so game and spunky and wanting to square himself. . . . To have him—such a Banner? Yes, dear; I believe it is the best way" (341). (If only all migrants were so pliable!)

Learning just how rigid is the class structure that defines his farm community, Ed Banner bows before the power of D. K. Starrett, "the big man of the community," a man whose influence extends from farmhouse to statehouse (189). Starrett even usurps Ed's parental authority in his own home. During the children's protest at the Banner housewarming, Starrett takes command after Ed orders the stubborn Ashley to drop the American flag he carries. When a third rancher counsels Ed that "D. K. usually has some pretty sound ideas," a chastened Ed shrugs and disappears into his office (189). Starrett countermands Ed later during similar family squabbles, first when Ed orders Mary home after she arrives at Valley Vista to observe the striking migrants (231) and then when Ed tries to stop her from speaking out in the mob scene after Emory's death (242). But Ed is well rewarded for his obedience. That he goes on to fight the "battles of agriculture" as a California state senator is not simple good fortune (317): "D. K. Starrett could manage anything, borrow anybody from anywhere" (237, 358). Starrett's power will now surely stretch from his Valley Vista Ranch to his state's capital.

Acutely aware of his place in the valley's pecking order, Ed, the future lawmaker, describes industrial farming's class structure to the press as completely natural. He instructs Kent Dexter, who is, appropriately, staying in the Banner home in a vacationing teacher's bedroom (246): "Starrett and Rugio—who have very big acreages and keep five hundred men or so all the year round, with a couple of thousand or more during the peak harvest seasons"—are in one "class." Ed's "class" consists of men who "work themselves and keep milkers and ranch-hands, and have extra help at harvest times" (291). The rest, "most of the men . . . have eight to eighty acres and do all the work themselves with the family's help, maybe hiring a few days' labor twice or three times a year" (291). At the bottom of the class pyramid are "regular ranch-hands," "extra men for the silo-filling," and, finally, migrants (6). But this caste system is visible only when farmers want it to be. For example, when Associated Farmers president Humphrey Pastor passionately calls for "right thinking" in America, he claims that growers will "clean house of the rats and the termites. America needs more than defenders of her liberties; she needs crusaders to reaffirm an American classless society" (295).

Not coincidentally, Pastor's words directly echo historical Associated Farmers president Holmes Bishop's. At the 1939 AF state convention Bishop claimed that "the greatest need in America today is right thinking" ("Farmers' Four" 1). Pastor's words also recall those Mitchell delivered at the same convention. Addressing the gathering, she called for a growers' "determination to clean house of the rats and the termites" ("Noted Authoress" 1). She declared that the migrant problem "calls for all we have of patience and courage and tolerance, of humor and determination and understanding, common sense, common humanity"—Pastor's words again—and she urged conventiongoers "to bring [California] back to the Union" ("Noted Authoress" 21).

## Cleaning House

After literally cleaning the Banners' new home, Mary and her farm neighbors dwell on their children's education. Critical of history teacher Pinky Emory, Hattie Westover suggests "marching down Main Street in a body with signs reading—'THIS IS NOT RED RUSSIA!'—'WE WANT OUR KIDS TAUGHT GOOD OLD AMERICANISM!'" (147). Hattie fears that Emory's teachings will harm "have-got[s]" like herself: listening to Emory, "you'll get an earful about what's going to happen to people that have sweated and toiled to get something and try to keep it!" (147). Taking the long view, these ranch women see the migrant problem as part of a national communist threat: "After all, there are lots like her. She's just a symbol of the sort of thing that's happening everywhere in our country" (148), a notion later picked up by Starrett (190). Vocal and visible reminders of fixed class distinctions in a United States that prides itself on equal opportunity, migrants/communists are a threat to the entire national entrepreneurial community. To safeguard itself through right-thinking education, the Banners' good old American community cleans house: the left-leaning Emory is fired (183).

Though its migrants are often dirty, *Of Human Kindness* itself is squeaky clean and Pollyannaish—no Banner sex, no foul language, no tragic endings. Mary sums up the novel's spotless and sappy optimism: "there's injustice and graft and greed here and everywhere, and ignorance and filth and suffering, but it's utterly false to say the world isn't better and going to be still better!" (178–79). When Ed suggests a farmworker's wife to care for his stricken mother, he recommends her only because she will be "a lot cheaper" and she "looks clean and not too dumb" (214). Unfortunately, this woman can only keep the kitchen "reasonably clean" (318). When nasty grower Joe Cosgrove, whose ill treatment of every living thing has given other growers a bad press, dies in a car accident, a farmworker describes him as a "dirty bum" who deserved to die (322). In response to Cosgrove's demise, Mary imagines death as "the sovereign dry-cleaner"—an odd remark that recalls Ed's disappointment that growers cannot "liquidate [migrants] by the thousands" (323, 95). Ashley's sometime girlfriend Nadine opens a beauty shop to teach women (farmwives and workers) how to be beautiful: "I can clean 'em up, powder and paint 'em up" (183, 215). Just as Mitchell does with class conflict, Nadine covers aesthetic differences with cosmetic solutions.

The novel may not describe Mary and Ed in bed, but it does suggest that growers must carefully monitor their workers' sexual conduct to save them from their own desire. Embodying communism's lure is the evil seductress, the Black Widow (206). At a roadside restaurant, Mary appraises this temptress in some detail, noticing that she wears a "thin white sweater which was drawn down revealingly over her round little breasts" (205). Having already seduced Ashley, this "Carmen—and Delilah—and Borgia" now entices a diner full of workers by "insinuat[ing] herself between two men who looked like cotton-pickers, putting red-taloned fingers on the shoulder of each, and leaning so low over one of them that the white wool over her breasts grazed his face" (206). In a voice "low, languid, faintly husky" she asks the racially mixed group of men: "how we coming? . . . Satisfied? *Like* to be the dirt under their feet, do you?" (206). Mary herself apparently is not immune to the Black Widow's wiles: "She's the most beautiful thing I've ever seen . . . and the wickedest!" (206). As if admitting the sexual and economic threat the woman poses, Mary, "trembling," drops her change, nickels and dimes, and notices that among the workers "no one offered to help her find them" (206). The Black Widow's voice then follows Mary outside, "sultry, persuasive" (207). If communism's siren song unnerves a sophisticated rancher's wife, what magic might it work on empty-headed migrants who covet the ranches?

By 1930 corporate America and the nation's education system had turned "cleanliness into a cultural value." Being clean was "patriotic, success-driven, and very American" (Hoy 123). In lockstep with this mainstream belief, *Of Human Kindness* approves of clean migrants because a clean worker is a profitable one. Helga Nielsen, a volunteer nurse working with migrants, preaches to growers the economic importance of migrant cleanliness: "it was economy in the end to teach them [the migrants] how to be clean, how to feed their children, how to prevent sicknesses" (327). Helga argues that Californians must "clean oop" farmworkers; "dirty children, sick children to-day, we must support tomorrow" (150). She tells Ed that the "migrant relief problem" is a "mess": "they're here, and they've got to be taken care of, some way. Can't just sweep the trash under the mat, like a lazy housekeeper" (95). To purify their community, growers must teach migrants "standards" of cleanliness: "How to keep their houses clean and their kids clean" (95). Serving up the Banners as decent examples, the novel often alludes to them keeping clean (63, 87, 176, 280). Nadine's Beauty Shoppe is Mary's refuge; she feels that her "safest" place is "under the drier after a shampoo" (298). As if safeguarding this "culture of cleanliness," the Banners board the high school's "head of the Domestic Science Department" (180; Hoy 152).

In this context, especially with farming increasingly defined as white-collar work, the frequent charge that Steinbeck's novel was filthy carried class-based overtones. Supporting burning and banning the book, W. B. Camp, president of the Kern County Associated Farmers, claimed: "We are angry . . . because we were attacked by a book obscene in the extreme sense of the word. . . . It is too filthy for you to point out its vile propaganda" (Kappel 212–13). The editor of the *Oklahoma City Times* claimed that the novel "'has *Tobacco Road* looking as pure as Charlotte Bronte, when it comes to obscene, vulgar, lewd, stable language'" (French, *Companion* 121). In a speech before Congress, Oklahoma congressman Lyle Boren defended his state, labeling the novel a "'dirty, lying, filthy manuscript'" (French, *Companion* 125). For such community leaders literature is no problem as long as it has nothing of substance to say. In Mitchell, these people would have found their literary idol: refusing to see obvious social ills, *Of Human Kindness* safely soils no reputations.

The Okies whom Mary encounters personally, the migrants who refuse to leave Banner and Westover property, are "more wretched than she had imagined" (344). They live under "torn and dirty pieces of canvas" (344); one woman is "an old grandmother . . . with the filthiest long claws" (343). Recalling Rose of Sharon's complaints, the new mother among them "pulled the soiled and sleazy silk blouse away from her thin chest where the small blue breasts hung limply down" and cries: "Ain't even no free milk fo' my Angie, here, an' she a'wailin' hongry!" (346).

When Mary offers these squatters ten dollars and a free can of cow's milk, Mitchell challenges the final tableau in *The Grapes of Wrath*—Rose of Sharon giving her breast to a starving man—by replacing what is essentially a gift from one human being to another with a financial deal. A commercial dairy farmer's wife, Mary tries to convince these miserable migrants to leave Banner property by bribing them, and she sees no problem in telling them not to drink the milk where they are: "Drive off on the main road, just a little way, first" (346). But before they leave, the Black Widow arrives and deliberately dumps the milk, "th' milk a' human kin'ness," to keep the migrants from listening to Mary (348). When the new, "crazy" mother attacks the Black Widow for spilling the milk, her "dirty nails rak[ing] the mother-of-pearl cheeks," Mary simply turns her back on both and leaves, essentially washing her hands of the whole incident (348). The message: good growers have done all they can to help these dirty migrants, but nasty communists have overturned their charitable efforts.

In its espousal of the entrepreneurial farm community, *Of Human Kindness* is shot through with odd contradictions. For example, though Mary's home is christened the "Community House," its membership is restricted to those who agree with the Associated Farmers (146). This is clear when, after their militarylike cleaning of the house (145), a commonwealth of like-minded farm women divides humanity into the "'have-got[s]'" and the "'have-nots'" and are fearful that have-nots will "just rush in and push [Mary] out" (147). Inconsistent in recalling her own AF convention speech, Mitchell describes the women as working "all day *like termites* to clean" house; these neighbor women even lay claim to the Banner place: "They were just staking out a claim. . . . This house was partly theirs now, for they'd squatted on it!" (147, 146–47; emphasis added). These inexplicable descriptions make Mary's neighbors akin to the migrant "squatters" she evicts from Banner land (344). One supposes that Mary's charity begins and ends at home.

The oddest contradiction points ahead to chapter 4. The novel betrays a peculiar nostalgia for the Old South: an allusion to "The Battle Hymn of the Republic" approvingly likens Mitchell's California farmers to antebellum slaveholders. Sung to the tune of "John Brown's Body," the "Battle Hymn" gave *The Grapes of Wrath* its name (Benson 387–88, 395).[9] At the housewarming Ashley and friends sing the song to protest Pinky Emory's dismissal—in what his father describes as a "revolt against authority" (187, 190). D. K. Starrett quickly steps in to reason with the student protesters solely to keep their parents from making "martyrs" of them (190). Wanting no public sacrifices to principles of fair treatment, Starrett lectures to the adults that martyrs are "pretty big people in the pages of history—our own history, too—and pretty hard to kill, because they don't stay dead! John Brown's body, for instance" (190). John Brown, of course, was the antislavery advocate who seized a federal arsenal at Harper's Ferry, Virginia, to incite a slave rebellion in 1859. His subsequent trial and execution galvanized antislavery sentiment in the North just before the Civil War. For Starrett, "The Battle Hymn of the Republic" is "dynamite" in its "concern for the underdog" (190), a phrase Ashley later uses to define communism: "the whole idea of helping the underdog" (223–24). But the Associated Farmers' concern is not for elevating the underdog, obviously, but with keeping the top dog on top. As Starrett might hope to do with migrants, he convinces the assertive children to know their place, and they soon come "downstairs without the signs and ready to go back to school"—where they will be taught the proper American way by approved teachers, such as those living with the Banners (191). As a telling footnote, Pinky Emory is killed right after she claims that the Associated Farmers have "enslaved" migrants (233).

Throughout, *Of Human Kindness* longs for the supposed genteel social relations of the antebellum South. At the scene of a car accident, Mary promises to care for the wife of an injured black man who recognizes her as "southern quality" (175). Keeping a hand on the dead woman until the man disappears in an ambulance, Mary then "went into the restroom and washed her hand[s]" of the whole incident (176). Asked by Ashley how the man knew that she was "southern quality," Mary notes that he was "the old-fashioned type, and I suppose he always associates decent treatment with the old ideal" (178). About the accident (two people die), Mary can only observe: "A thing like this gives us pause, doesn't it? Makes us take account of stock, stand off and look at ourselves" (178). Viewed from her Associated Farmers' perspective, Mary sanctions the marriage of southern quality to "California common sense and common humanity," thus replanting antebellum southern paternalism in West Coast farm fields (178).

Underscoring the novel's nostalgia is the fact that the Ashley family wealth originated with pre–Civil War Tennessee plantations (47). Mary's cousin Isabelle even holds an office in the Daughters of the Confederacy (9). Banker Burton Doane, whose family owned plantations near the Ashleys, now lives with his father "in the grave old house with the three Negro servants who were the third generation of the ones who had come with the family from Tennessee. It was Mary's private conviction that even the young ones were unaware of the fact that one Mr. Lincoln had lived and died in their cause" (104). The most horrifying thing about this yearning for old-time plantation labor relations is that Mitchell offers no alternative to suggest that she is being ironic.

But then again, one reason that *The Grapes of Wrath* was a best-seller was that it portrayed whites being treated as if they were not white. Until the 1930s migrant exploitation did not trouble the national consciousness because before that mainly racially mixed, foreign-born labor worked the fields; Chinese, Japanese, Filipinos, and Mexicans all dominated California farm labor at different periods. In 1939 Carey McWilliams noted: "The established pattern has been somewhat as follows: to bring in successive minority groups; to exploit them until the advantages of exploitation have been exhausted; and then to expel them in favor of more readily exploitable material" (*Factories* 305–6). Each group worked generally unseen by the mainstream media and the general public.

With the influx into California of Dust Bowl refugees, who were depicted overwhelmingly as native, white American farmers in news reports and FSA photographs, the problem soon became the crisis it really always had been—only this time, the problem generated a lot of press. McWilliams notes that "it was suddenly realized in 1937 that the bulk of the State's migratory workers were white Americans and that foreign racial groups were no longer a dominant factor." "These despised 'Okies' and 'Texicans' were not another minority alien racial group (although they were treated as such) but American citizens familiar with the usages of democracy" (*Factories* 305–6). With racial stereotyping now unavailable to them as a strategy to distinguish themselves from most migrants, growers turned to demonizing communists and defining the newest migrants as unclean and lazy.

The United States's entry into World War II signaled the end of the *white* migrant labor problem in California. It did not end the migrant problem, however. Carey McWilliams's belief that the "influx of thousands of transients" had created "a superabundance of skilled agricultural labor" that could solve the problem proved to be shortsighted, especially his assertion that "the race problem has, in effect, been largely eliminated" by this abundance (*Factories* 324). But with white workers moving into the armed forces and the defense-related industries spawned by the war, a gap soon developed in a California agriculture that still relied on seasonal labor. That gap was filled after the war by a government-sanctioned bracero program, which brought thousands of Mexican workers into California fields. Consciously a question of class for Steinbeck and Mitchell, the industrialization of agriculture quickly becomes an issue of race for others, especially when one turns to Luis Valdez's Chicano theater *actos* (1965) and African American works such as Ernest Gaines's *A Gathering of Old Men* (1983). Not thirty years after the Steinbeck/Mitchell debate, supporters of Chicano migrant labor leader Cesar Chavez were using *The Grapes of Wrath* to publicize a 1965 grape strike to white middle-class America (Benson 423).

Though *The Grapes of Wrath* has earned an enduring place in American literature, the values represented in *Of Human Kindness* have won the larger economic and cultural battles. The farm community providing our food and fiber today is entrepreneurial; the yeoman community survives only in isolated pockets, such as with the Amish and in our collective memory. Statistics illustrate the extent of the entrepreneurial ideal's chokehold on U.S. agriculture. For example, the average American farm increased in size 100 percent between 1945 and 1985 (Summerville 28). Between 1960 and 1975, 45 percent of American farms disappeared, a loss in agricultural land of six million acres, roughly the size of Maryland (Summerville 26). The number of farmers dropped from 3 percent of the population in 1985 to

1.9 percent in 1993 (Summerville 28; Vobejda A1). The U.S. Census Bureau stopped counting farmers in 1993 because "an increasingly large minority of farmers are divorced from their workplace, as much as typically urban workers are from theirs. . . . The farm as the homestead seems to have less cultural validity than it once did" (Vobejda A13). The entrepreneurial model oversees food production: "nearly a third of farm managers and 86 percent of farm workers live away from the farm and commute to the fields" (Vobejda A13). The number of farm operators who worked two hundred days or more off the farm increased from 665,570 in 1992 to 709,279 in 1997 ("Highlights"). This kind of development is, after all, how we define progress. But is it worth it?

[F]ire at that tractor.—Ernest Gaines, *A Gathering of Old Men*, 1983

CHAPTER FOUR

# Racism and Industrial Farming

## *Actos* (1965) and *A Gathering of Old Men* (1983)

On 22 April 1993 United Farm Workers Union (UFW) president Cesar Chavez died in his sleep while on union business in San Luis, Arizona (Matthiessen). A leader of the successful 1965–70 table grape boycott that led to growers' acceptance of the United Farm Workers Organizing Committee (UFWOC), Chavez fought long and hard for Mexican American fieldworkers' rights and recognition. The union he helped to found first gained national attention in September 1965, when it joined Filipino workers on the picket lines against San Joaquin grape growers near Delano, California. From the subsequent strike was born a Chicano civil rights movement that was fueled in large part by an energetic theater that defined Chicano ethnic solidarity in political and cultural terms.

Though widely admired for his work, including praise from presidents and popes, Chavez was not universally liked, especially among the growers with whom he struggled to make the union legitimate. The grower attitude noted in chapter 3—that farmworkers were to disappear between harvests—was still the norm in the 1960s; Mexican American workers were not considered full members of a predominantly Anglo-owned California agriculture. The tensions that the strike generated can even be detected in obituaries of Chavez and Julio Gallo, the "co-founder of the world's largest winery" and one of Chavez's antagonists in a strike and wine boycott begun in 1973 ("Julio"; Levy 495).

The notices appear on the same page in the June 1993 issue of the pro-grower *California Farmer*. In contrast to the Gallo obituary, Chavez's subordinates his accomplishments to criticism of his efforts. Matter-of-fact, though respectful, the obituary points out that he was a "champion of the rights of migrant farm workers . . . the first farm-labor organizer . . . noted for his use of boycotts and fasts to publicize his causes." But claiming that he was the first farm labor organizer suggests that such activity is recent, when in fact it has a long history. Union organizers had operated for decades among migrant workers in California fields: in 1928, for example, the Confederación de Uniones de Obreros Mexicanos was founded in southern California to "promote bread-and-butter unionism among Mexican workers," and communist organizers were active in farmworker unionization in the Imperial Valley in the 1930s (Daniel 106–8, 112–17). Denying a long and acrimonious history of farm-labor relations implies that Chavez is the origin of all farm labor trouble, something big grower readers might well believe.

Chavez's obituary carefully confines his achievements to phrases, whereas independent clauses focus on the labor leader's faults: "As the founder and only president of UFW in its 31-year history, Chavez had drawn criticism for his rigid control of the union, distrust of dissent and refusal to listen to opinions contrary to his own." De-emphasizing his organization of farm laborers and downplaying his position as president, the sentence condemns his tenure as UFW leader, using such negatives as "rigid," "distrust," "refusal," and "contrary" to mark Chavez as intransigent and self-centered. The obituary identifies him as the sole initiator and benefactor of collective, union activity: "his use," "his causes," "opinions contrary to his own."

The Gallo obituary, however, attributes negative information about the man to the company. For example, a long dependent clause harbors an embarrassing note regarding the origin of Gallo wealth. That the Gallo company made its money in the "'misery market'" appears in a subordinate "while" clause, but the sentence's independent clause implies that such sins are forgiven because "the company has in recent years successfully entered the premium wine market." In this third paragraph, the obituary's longest, Gallo the man appears only once as a subject: "He also pushed the company toward organic farming practices." The paragraph's trajectory follows Gallo as man/company from rags to farm riches, moving as it/they do from filling misery market brown bags to gracing wealthy connoisseurs' dining tables.

In conflating its subject, Gallo the "farmer," with Gallo the company, the obituary assumes a fully industrialized agriculture. Though the farmer is dead, the incorporated Gallo lives on: after all, "Gallo now enjoys an 18-percent market share in the varietal wine business." Though the following sentence's pronoun must refer to the man, its antecedent is Gallo the company: "He also pushed the company toward organic farming practices, and his death came before another prized project could be completed: release of Gallo's high-priced, organically produced chardonnay and cabernet sauvignon." The accomplishments of this double Gallo—farmer and corporation—are underscored and held up for readers' edification by the obituary's word choice: "most recognizable," "astounding," "successfully," "fulfilled a dream," "enjoys," "prized project." Chavez's obituary is silent about union accomplishments.

Though the Chavez obituary minimizes the man's achievements, it took something so simple as the very appearance of the notices on the same page to inflame at least one letter writer's passions. The names' alphabetical ordering may appear innocuous, but it was picked up by an angry Frank Light of La Jolla, California, in the August issue of *California Farmer*: "I cannot believe that the *California Farmer* would have the nerve to print the obituaries of Julio Gallo and Cesar Chavez in the same publication, let alone on the same page." Light's inversion of the magazine's alphabetizing registers the depth of feeling that persists among some California residents regarding Chavez's legacy. His reordering of names suggests, too, a restoration of industrial agriculture's hierarchy of management and labor.

But Light reveals his greatest distress when he demands that *California Farmer* should have featured Gallo on its "front cover as a hero of California farming." Light's support for this claim lies in the fact that Gallo "revolutionized the marketing and production of California wines." But this revolution began, as Gallo's obituary notes, with "cheap, fortified wines for the inner-city 'misery market.'" Light apparently has no trouble applauding the fact that Gallo's revolution in marketing was based on the exploitation of the poor, not to mention the denial of migrants' rights at the production end. Though he does not explicitly touch upon these populations' problems, Light does not ignore them. His answer to the poor and to Chavez's Mexican Americans is implicit in his celebration of Gallo as a man who "proved that the 'American Dream' of success through hard work, within the rules of our democratic capitalistic system, can come true." Ignoring, or unaware of, the inherent contradictions in the phrase "our democratic capitalistic system," Light suggests that Chavez's unionism has no place in such an order: "Cesar Chavez, in contrast, wanted to milk the blood from the veins of our free enterprise system without paying for it." In the logic of hard-core capitalism, draining blood from an economic body is, apparently, perfectly fine—as long as one pays in cash. The body's health is secondary.

Even those who support Chavez catch themselves in a verbal web that denies farmworkers recognition as a vital part of California farming. Mainstream press reports of his death, which were generally sympathetic to his work, claim that in his efforts to unionize Mexican American fieldworkers Chavez was "a David taking on the Goliaths of agriculture" (Lindsey 29) and "an unlikely David to go up against the four-billion-dollar Goliath of California agribusiness" (Matthiessen). As heroic as it sounds, the David-and-Goliath analogy does Chavez no service. When writers like Peter Matthiessen identify Goliath *as* agriculture, they place Chicano fieldworkers *outside* of agriculture, thus denying them their rightful place as participants in California farming. And this, ironically, was just what growers hoped for. But at least the hyperbole in the David-and-Goliath analogy registers how far American agriculture has moved from mid-nineteenth-century perceptions of the hired hand as family to today's wide divide between farm labor and management.[1]

## Strike Theater

The Delano strike, the first major grassroots protest against industrial farming since the 1930s, pitted a community of farmworkers against individual growers, among whom were two absentee landowners with vast acreage: DiGiorgio Fruit Corporation, whose DiGiorgio Farms managed nearly 4,400 acres in Delano, and Schenley Industries, Inc., which operated 3,500 acres (Dunne 127). Striking farmworkers demanded better wages, improved working conditions, and recognition of the union as their bargaining representative. Strict pesticide control became an official strike issue in 1969 ("Boycott Unit"; Roberts, "2 Big Issues"; see "Chavez Scores"). To achieve its early goals, the union began a nationwide boycott against DiGiorgio and Schenley products, publicized with a Boston Grape Party and, more significantly, a pilgrimage to California's state capital in March 1966, accompanied by the newly formed Teatro Campesino, or Farmworkers Theater (Dunne 129, 131–33). The boycott soon forced Schenley to the bargaining table, and the company signed an agreement with the union on 6 April 1966 (Dunne 136). One grower described the union's boycott as "unparalleled in American history, literally clos[ing] Boston, New York, Philadelphia, Chicago, Detroit, Montreal, Toronto completely from handling table grapes" (Levy 296).

But DiGiorgio held out, shielding itself from the farmworkers' union by courting the Teamsters Union, which had up to that point been a vigorous supporter of the National Farmworkers Union (Dunne 141). Though on 24 June 1966 some DiGiorgio farmworkers voted for Teamster representation, Chavez claimed victory because an election boycott kept many workers from casting ballots (Dunne 145). Needing "money and organizing strength for his battle with the Teamsters," the farmworkers' union merged with the AFL-CIO to become the United Farm Workers Organizing Committee (Dunne 155). The UFWOC won a supervised second election on 30 August (Levy 243–46; Dunne 166). The next year DiGiorgio caved in and signed union contracts ("Fruit Concern"; Dunne 168). Though Schenley and DiGiorgio capitulated, the rest of Delano's growers adamantly refused to negotiate until June 1969; negotiations dragged on until finally twenty-six Delano growers signed a collective bargaining agreement with the union in July 1970 (Day 166). Following its victory among grape growers, the union turned its attention to lettuce growers, who had also run to sign with the Teamsters. Again, strikes and boycotts were used to get growers to negotiate (Levy 327).

According to news reports occasioned by Chavez's death, the labor organizer had only stunned his opponents: California agribusiness absorbed the union's strikes with little change to its fundamental workings.[2] Though the UFW enjoyed early success, it has not taken hold as Chavez had envisioned, and its future, "already in doubt with the influx of nonunionized, illegal immigrants . . . is now less secure than ever" (Gollner). Peter Matthiessen attributes the failure of the present-day union to the fact that it lacks

> the fervor of those exhilarating [1960s] marches under union flags, the fasts, the singing, and the chanting—*"Viva la huelga!"*—that put the fear of God in the rich farm owners of California. These brilliant tactics remained tied in the public perception to La Causa, a labor and civil-rights movement with religious overtones which rose to prominence in the feverish tumult of the sixties; as a mature A.F.L.-C.I.O. union, the U.F.W. lost much of its symbolic power. Membership has now declined to about one-fifth of its peak of a hundred thousand.

The Delano strike sparked a new, politically charged era of Mexican American self-awareness, which in turn fanned a high energy Chicano movement that burst out of California's fields to sweep the nation. The symbolic power that Matthiessen claims the union lost lay in its very real and very close ties to the community it served. A key part of that symbolic, grassroots appeal was a reinvigorated Chicano theater: the grape strike witnessed, and was witnessed by, the birth of perhaps the most conspicuous connection between agriculture and literature—Teatro Campesino (Farmworkers Theater) and the new genre it created, the *acto*.[3]

*Actos* were communal responses to injustices inflicted upon Chicanos by California's industrial agriculture—low wages, poor working conditions, racial discrimination. According to Luis Valdez, founder of Teatro Campesino, "the acto is Chicano theatre" (*Early Works* 11). A short skit, each acto was usually fifteen minutes long, performed with a cast of two to three people, using "no scenery, no scripts, and no curtain" (Valdez, "Beginnings" 115). Props and costumes were simple, including signs to identify what actors represented; actos were performed almost anywhere—in a home or on a flatbed truck along a picket line (Valdez, "Beginnings" 115). Not written, actos were "created collectively, through improvisation by a group" (*Early Works* 13). Valdez writes: "each [acto] is intended to make at least one specific point about the strike, but improvisations during each performance sharpen, alter, or embellish the original idea" ("Beginnings" 115). The line between audience and actor hardly existed; audience participation was expected, virtually required, for an acto to be successful: "*Audience participation is no cute production trick with us; it is a pre-established, preassumed privilege*" (Bassnett-McGuire 19). Because the acto addressed a specific audience of Chicano farmworkers who shared life experiences and a common culture, satire was quite naturally its most effective weapon.

Actos' community performance validated the individual farmworker's experience, creating a shared vision of what the strike was about and who exactly the strikers were. Its task was political and rather dangerous: "Teatro Campesino found itself performing in the face of the 'enemy,' poised on its flatbed truck within earshot (gunshot) of local gendarmes eager to cause a little destruction. Members of the troupe . . . recall how . . . the term 'stagefright' took on a new meaning in the face of goon squads and fully garbed riot police" (Huerta 64). Striking farmworkers not only performed the actos, they created them: "Starting from scratch with a real-life incident, character, or idea, everybody in the Teatro contributes to the development of an *acto*" (Valdez, "Beginnings" 115). Valdez declares that the acto's "major emphasis" was a "social vision, as opposed to the individual artist or playwright's vision" (*Early Works* 12–13). The actos' "archetypes symbolize the desire for unity and group identity" among Chicanos fighting growers who championed only individual competition (*Early Works* 13).

Teatro Campesino's close connection to the people it served is best illustrated in Valdez's description of the moment when the idea of the acto first coalesced—at a small gathering of student strike volunteers and farmworkers where he tried to explain his ideas for a theater group:

> I talked for about ten minutes, and then realized that talking wasn't going to accomplish anything. The thing to do was do it, so I called three of them [farmworkers] over, and on two hung *Huelgista* signs. Then I gave one an *Esquirol* sign, and told him to stand up there and act like an *Esquirol*—a scab. He didn't want to at first, because it was a dirty word at that time, but he did it in good spirits. Then the two *huelgistas* started shouting at him, and everybody started cracking up. All of a sudden, people started coming into the pink house from I don't know where; they filled up the whole kitchen. We started changing signs around and people started volunteering, "Let me play so and so," "Look this is what I did." . . . The effects we achieved that night were fantastic, because people were acting out *real things*. (Bagby 74–75)

Though the "real things" Teatro Campesino focused on included class divisions similar to those represented in Ruth Comfort Mitchell's *Of Human Kindness* and John Steinbeck's *The Grapes of Wrath*, the theater also examined the racial divide between grower and worker in representing the vehement antilabor, anti-Chicano stance of California farmers during Chavez's unionizing attempts. Linking race and industrial farming, Valdez claims that Chicano culture offers others an answer to "the American power structure. . . . A perspective and a way of life that does not include the systematic genocide of unwanted races or a technology that is being used only to destroy the very earth that mothered us" (*Early Works* 16). To celebrate La Raza, Teatro Campesino defined Chicano identity as unique and distinct from that of Anglo growers, tracing as it did Chicano origins back to roots among Aztec and other Native American groups.

The actos and the Delano strike were born in an atmosphere poisoned by a long history of racial discrimination and distrust. Chavez himself observed: "Here in the fields, for ages the employers worked the races one against the other in competition for jobs. That was a tactic to keep wages down and keep unions out" (Levy 198). Incidents of "subliminal humiliations" such as those in the San Joaquin Valley's educational system were common: at "a Valley elementary school where the student body was fifty-eight percent Mexican. . . . One teacher, asked why she had called on an Anglo boy to lead five Mexicans in orderly file out of the classroom, replied, 'His father owns one of the big farms in the area and one day he will have to know how to handle Mexicans'" (Dunne 63). The Delano strike was carried on in this racially divisive climate; "innuendo of racial discrimination creeps into every strike discussion on the West Side [of Highway 99, where the National Farmworkers Association headquarters was located]" (Dunne 60).[4]

In his on-site observations of the strike, John Gregory Dunne notes: "The term 'Anglo,' as it is used in the Valley, encompasses all whites, save Mexicans, whether they be Christian or Jew, of eastern or southern European descent" (10). At Delano, Anglos of mainly Yugoslavian ancestry owned the largest grape farms, and laborers of mainly Mexican and Filipino extraction worked them (Dunne 10). Teatro Campesino responded to racial tension in the fields through several actos, including *Las Dos Caras del Patroncito* (1965) and *Vietnam Campesino* (1970).[5]

## The Two Faces of the Boss

The first published acto, *Las Dos Caras del Patroncito*, was first performed on the Delano picket line in 1965 (Hernández 34). The acto has three characters: an Esquirol, a farmworker scab; a two-faced Patroncito, a grower; and Charlie, an apelike armed guard. Its action centers on a role reversal between the Esquirol and the Patroncito. As Guillermo Hernández points out, the exchange is "a symbolic representation of fate's whimsical assignment of social roles" (34). Though illustrating how "the fortuitous circumstances of birth" dictate who is a rich grower and who is a poor farmworker, the acto exposes the pastoral mask that corporate agriculture hides behind and broaches the specter of racial discrimination (Hernández 35).

According to Valdez, *Las Dos Caras del Patroncito* was created in response to the "phoney 'scary' front of the ranchers," put on to intimidate strikers; the acto "was intended to show the 'two faces of the boss'" (*Early Works* 17–18). Critics like Hernández rightly point out that the grower's two faces represent projections of his arrogance while wielding power and his rebellion when he receives unjust treatment (35). But the crux of the acto and the origin of the Patroncito's undoing lie in the acto's satire of the Patroncito's two-faced view of himself, a self-image created by the Anglo industrial agriculture he represents—and that Valdez's audience was fighting with for recognition. In showing his first face, the Patroncito describes himself as "an important man, boy! Bank of America, University of California, Safeway stores, I got a hand in all of 'em" (19). All these—the bank, the university, and the stores—represent the pillars of advanced industrial agriculture, more particularly its tendency toward vertical integration: heavy capital investment, research/development, and distribution, not to mention production. The gluttony implicit in this total control of a food system is symbolized by the "*yellow pig face mask*" the Patroncito wears (18). The whip he carries and his denigration of the farmworker as "boy" suggest that he sees his modern-day farm as a pre–Civil War plantation (23).

But the grower hides behind another face: the face of a poor yeoman, symbolized by his unshined shoes. He points to his shoes to divorce himself from his agribusiness image: "But look, I don't even have my shoes shined" (19). That he tries to stop the farmworker from shining those shoes only serves to illustrate how wedded he is to the image of being a farmer "in the field" (19). As if to protect this image, his armed guard shows up and "*lunges for the* FARMWORKER" to stop him (19). Later, the grower again invokes this view of himself as a poor, hard-working farmer victimized by "commie bastards": "Who the hell do they think planted all them vines with his own bare hands? Working from sun up to sunset!" (22). Though forced to admit that all of this describes his grandfather, he still contends that he "inherited it, and it's all [his]!" (22). What he has inherited, of course, is not only his grandfather's land and money but also his grandfather's agrarian self-image, an image that is in fact just that—an image. The contradiction in wearing two masks—industrial farmer and yeoman farmer—leads the Patroncito to imagine a third identity for himself: Chicano laborer. His inability to center on one identity, as the farmworker finally does, illustrates his impotence in the face of those who can, a powerful message for a striking farmworker audience.

By this time, the grower has already recounted his "problems" to the farmworker to show him how tough it is to be a poor farmer (21). The grower "gotta pay for what [he] got," including his $12,000 Lincoln Continental, his $350,000 ranch house, his wife's $5,000 "mink bikini" and her weekend trips "to L.A., San Francisco, Chicago, New York" (21–22). He ends his catalog of problems with a lament: "Me, all I got is the woman, the house, the hill, the land" (22). The lucky farmworker, on the other hand, lives in "nice, rent-free cabins, air-conditioned," eats for free, rides to work for free; in short, he has nothing "to complain about" (20–21). The white grower's blindness to the circumstances of his Chicano workers as he reels off his high-priced problems is conditioned by his fantasy that those workers live a pastoral life and not, as they do, the life of exploited laborers in a highly industrialized and racially hierarchized agricultural system.

This myth of farmwork's inherent pastoralism is the complement of the myth of the hard-working yeoman. That the grower refuses to see the reality behind either myth is apparent when he remarks, without irony, that he sits in his office wishing to become a field hand: "Sometimes . . . I wish I was a Mexican. . . . Riding in the trucks, hair flying in the wind, feeling all that freedom, coming out here to the fields, working under the green vines, smoking a cigarette, my hands in the cool soft earth, underneath the blue skies, with white clouds drifting by" (23). What else could a man want? When the farmworker shouts "I want more money!" the grower cannot believe that he would trade his pastoral position for the grower's heavily capitalized one: "Shut up! You want my problems, is that it? After all I explained to you?" (23). But then, in an ultimate show of corporate control, the grower claims that he has the "power" to transform identities, to make his and the farmworker's dreams come true; he can demote himself to farmworker and promote the farmworker to grower (23). Unfortunately for the grower, the farmworker warms to his new role and the Patroncito learns that farmworkers do not lead easy, pastoral lives.[6]

The acto ends by suggesting that racism has been an implicit element throughout. After the Patroncito-dressed-as-farmworker accuses the farmworker of raping his "white wife," the armed guard, Charlie, turns on the Patroncito, "*infuriated*" by the charge, shouting: "You touched a white woman, boy? . . . I'm gonna whup you good, boy!" (26). Unmindful of the Patroncito's cries about his real identity, Charlie carries him away to discipline him for his attempted sexual transgression of racial lines. As he is dragged off, the Patroncito turns to the community for help, shouting: "Somebody help me! Where's those damn union organizers? Where's Cesar Chavez? Help! Huelga! Huelgaaaaa!" (27).

In the acto's final lines, the scab farmworker declares his unity with the strikers by refusing to retain the Patroncito's identity. He strips off the pig mask and facing the audience declares that he will only keep the Patroncito's cigar—a representation of the grower's power, both social and sexual. Nor will he keep the grower's house, hill, or land. Retaining the cigar asserts his freedom after enduring the indignities of shining the Patroncito's shoes (19), acting like a "*docile dog*," and "*kiss[ing] [the grower's] ass*" (20). In empowering himself, the farmworker reinforces for audience members that they are not striking to invert the fields' labor hierarchy: they stand together for justice and to assert a distinct Chicano identity.

### *Vietnam Campesino*

Initiating Teatro Campesino's transition from a farm labor focus to an interrogation of other aspects of Chicano experience, *Vietnam Campesino* (1970) appeared within the local circumstances of a lettuce strike and the national context of an unpopular Vietnam War (E. G. Brown 30). In July 1970, when lettuce growers in the Salinas Valley signed agreements with the Teamsters to represent their workers, the farmworkers' union rushed to sign others to its own ranks, and the battle to win workers was on (Levy 329). The day after a 16 September 1970 court decision granted "permanent injunctions against picketing to thirty growers," Chavez's UFWOC announced a lettuce boycott (Levy 422). In the midst of all this, and throughout the grape strike, the war in Vietnam raged.

The racial polarization forming the backdrop of *Dos Caras* comes to the forefront in *Vietnam Campesino*, first performed, not coincidentally, at "Guadalupe Church in Delano during the annual Thanksgiving gathering of huelguistas and UFWOC supporters" (*Early Works* 98). The acto satirically looks at how wealthy growers and the military, the "*military-agricultural complex*" (98), symbolized in the figures of General Defense and Butt Anglo, collude in a brutal, systematic repression of Chicano farmworkers and Vietnamese peasants in the name of "Fighting Communism" and saving "the crop from Communism" (119).

The acto's opening illustrates that military and agricultural interests are complementary. Butt and the General disperse picketers with emblems of the law and order that they impose throughout to maintain their "Agribusiness" relationship (103): Butt wields an American flag (99) and the General fires his gun (100). After scattering the picketers, they spot each other and join forces for mutual protection—"We'd better get together on this!" (100). Visually underscoring their ties, Butt and the General walk arm in arm (100), shake hands (101), and scratch each others' back (102). Photographs of the acto's performance show the General wearing a pig face mask, similar to the one that the grower wears in *Dos Caras*.[7]

Much like the Patroncito in *Dos Caras*, Butt Anglo hides behind a yeoman farmer image to confront antiwar picketers: "Why are you yelling at me about the war in Vietnam? I'm just a poor grower" (99). A little later, he turns to the audience and asks: "How do you like that? Anti-war pickets against me, a simple dirt farmer" (99)—this after revealing in asides that he owns fifty thousand acres and receives a "couple of million" dollars in federal subsidies (99). Trailed by striking farmworkers when he first appears, the General cynically wonders why they are picketing him, only to be reminded that disproportionate numbers of Chicanos are dying in Vietnam and that the Pentagon buys scab grapes for consumption by American troops there (100).

The acto teaches its striking farmworker audience that their troubles are not unique—they are fighting a worldwide pattern of American military and agricultural aggression. While U.S. troops fought in Vietnam, American agricultural scientists were carrying forward the Green Revolution, introducing high-yield crop varieties, chemical fertilizers, and pesticides to Southeast Asian nations, effectively displacing farmers and indigenous farm practices (Shiva, "Masculinization" 112–13). In the acto, the connection between Chicano farmworker exploitation and the war in Vietnam is introduced visually when Vietnamese peasants and Chicano farmworkers

Photograph of *Vietnam Campesino*. From the Teatro Campesino Archives, California Ethnic and Multicultural Archives, Special Collections, Davidson Library, University of California, Santa Barbara.

appear on opposite sides of the stage and "BUTT ANGLO *and the* GENERAL *move between them for the rest of the acto*" (110). The striking Chicanos on stage are aware of their solidarity with the Vietnamese: "PADRE: (*To his wife.*) Oye, vieja, esas gentes son iguales que nosotros" (Listen, old lady, these folks are just like us) (115). But farmworkers, the Vietnam War, and the Green Revolution are linked most clearly by Butt's and the General's use of pesticides to eliminate both campesino families.

*Vietnam Campesino* warned farmworkers of the dangers of pesticide poisoning, which did not become an issue until the year before the acto was first performed. In March 1969 Chavez told the *New York Times* that "the real problem is that the workers don't even know how dangerous this stuff is" (Roberts, "Charge"). In January the union had expanded its grape boycott to "include a campaign against the 'wholesale use' of pesticides" ("Boycott Unit"). Chavez went so far as to say that he was "willing to set aside the issues of wages and working conditions if grower groups would negotiate the pesticide issue" ("Boycott Unit"). The union took pesticide-linked health problems seriously: "Because of the pressure of the UFWOC, the California State Department of Public Health conducted a survey on farm workers in Tulare County. The results showed that 80 per cent of the workers interviewed had suffered adverse effects because of their contacts with the poisons" (Day 93). When the union sought access to Kern County Agricultural Commission records "on how and where pesticides were used in the county," Commissioner C. Seldon Morley refused to provide it, hiding behind the claim that " 'Agricultural pesticides and chemicals are a way of American life today' " (Roberts, "Charge" 46).

Strikers bargained hard for pesticide control, and restrictions were part of the "*first major contract signed with the grape industry: the Coachella Valley contract with the David Freedman Company. It set the pattern for all subsequent agreements*" (Day 205). The contract's section XVIII, part B, bans the use of "DDT, Aldrin, Dieldrin, Endrin, Parathion, TEPP and other economic poisons which are extremely dangerous to farm workers, consumers, and the environment" (Day 214). Naming pesticides "economic poisons" was a farmworker victory because the phrase stresses the role of pesticides as moneymakers in industrial farming, rather than as pest controllers. At the same time, the U.S. military was using similar poisons, such as Agent Orange, to defoliate Vietnam.[8]

Dow Chemical Corporation manufactured the poison that killed Chicano and Vietnamese campesinos (101, 117). A maker of herbicides and pesticides for agriculture, Dow also manufactured key ingredients of Agent Orange for military purposes (Uhl and Ensign 123). In agriculture, of course, herbicides and pesticides eliminate pests to ensure survival of a grower's cash crop. Hence, the acto draws a parallel between the worker/striker and the pest—both must be exterminated to produce a successful crop to keep the grower in business. Not incidentally, Agent Orange was derived from a post–World War II cotton defoliant that was created to assist machine harvesting.[9] Militarily, herbicides, such as Agent Orange, defoliated the Vietnamese landscape, poisoning peasant farms—and American soldiers—to expose a military pest, the Viet Cong (Uhl and Ensign 123).

Throughout the acto, the audience sees pesticide poisoning inflicted in only one direction, from whites to Chicanos and Vietnamese. For example, Butt's son, Little Butt, sprays labor contractor Don Coyote to "have a little fun with the spic," quickly blinding him (104). Later, as an air force flyer, Little Butt dive-bombs both Chicanos and Vietnamese campesinos with death-dealing pesticide-laden lettuce (118). Stage directions in the dive-bombing scene call for showing, "*Depending on the circumstances, slides of Vietnam, farm labor, crop dusting, dead bodies, etc.*" (118). These moments recall incidents during the Delano strike when growers turned their sulfur sprayers on pickets (Levy 189, 194; Dunne 131). Underscoring the connection between agriculture and the military, General Defense remarks, "You spray pesticiedes [*sic*], and I bomb Vietnam" (119). He asserts this with a racial hierarchy in mind: he notes that the American public has been deceived by his plan in Vietnam "for the last ten years. You get the Whites to pay for it, and the Chicanos, the Blacks, Indians and Orientals to fight it. . . . So what if we've killed the people? We've saved the country from Communism" (119).

The suggestion that food is a weapon alludes to a serious debate at the highest levels of the U.S. government, one still unsettled. Earl Butz, Richard Nixon's secretary of agriculture, coined the phrase " 'food as a weapon' " and "subsequent authors have clearly intended that food trade be analyzed on a par with tanks and missiles, as a component of the U.S. arsenal" (Thompson 30–31). In response to the Soviet invasion of Afghanistan in 1979, for example, President Jimmy Carter ordered a grain embargo against the Soviet Union (T. Smith 1). On 23 December 1980 Ronald Reagan's secretary of agriculture John Block asserted that "food was America's greatest foreign policy weapon . . . 'I believe food is now the greatest weapon we have for keeping peace in the world' " (King). Virgil's *Georgics* metaphor—farming is warfare—is now a cold-blooded fact of international politics (Perkell 36). If food is officially a weapon, we are all, always, combatants in a deadly world war.

Like growers who want to keep pesticides off the bargaining table, whites in the acto conceal pesticide spraying: "LITTLE BUTT *hides airplane behind his back*" (107). The elder Butt denies pesticide dangers when he describes their smell as "country smog," though he chokes on his own words (106). His allusion to smog, a predominantly urban phenomenon, suggests that his vision of agriculture is an urban one, implying his divorce from the land he controls and the workers he uses. Attempting to make pesticides palatable, he links the poisons with processed food, claiming that their smell is "just a few chlorinated hydrocarbons, mixed in with some organo phosphates. Sounds like a new breakfast food," which he encourages the farmworkers to eat (106).

The acto also warns its Chicano audience of the poison of patriotism. While Little Butt escapes to college to join the Air Force ROTC, Hijo is drafted to fight in Vietnam. As a soldier, he is ordered by General Defense "to burn the house of these farmworkers" (116). The Chicano soldier mistakenly goes for his parents' house first, only to refuse to destroy it when he realizes whose it is. Tragically, he then obeys orders to burn a similar home, the Vietnamese one, and is killed for it by a Vietnamese peasant, all this under the blank gaze of his Anglo commander (116). The vision of a young Chicano, unwillingly separated from his family in the first place, destroying a "*small labor camp shack*" would have had powerful resonance with an audience of farmworkers who place a high value on family and community, especially given the acto's original performance site and time—in a church on Thanksgiving Day during the unpopular Vietnam War.

That the General posthumously names Hijo a hero is only one example of the acto's attention to how Anglos impose identities on Chicanos. In the acto, Anglos assign identities to Chicanos and the Vietnamese to justify eliminating them; General Defense demonizes both groups with "good ugly sounding name[s] . . . name[s] you want to kill, you want to destroy!" (111). Thus the Vietnamese are "gooks! They aren't people, they're gooks" (111). The Anglo assigning of names/identities to Chicanos during the grape strike—Communists, Viet Cong, for example—is recalled when the General assigns identities to the campesinos as part of his "public relations" plan to help Butt fight an antipesticide campaign. His plan is similar to the one he has been successfully using in Southeast Asia: "And after ten years in Vietnam, I oughta know" (110).

Key to the General's plan to restrain Chicano strikers is the negative renaming of their leadership. Just as members of the Vietnamese National Liberation Front were renamed the Viet Cong, United Farm Workers Organizing Committee leaders Cesar Chavez and Larry Itliong must be renamed the "Communist Mexican Chong" (111). The second part of the General's plan, which culminates in Little Butt's lettuce bombing, also involves imposing an identity on Chicanos. After failing to get the Chicanos to join the growers' union, Butt Anglo, the General, and Don Coyote huddle and decide on the Teamsters as a good phony front union (113). Though the campesinos refuse to join, the General claims that all the three need to do is to force a name on the strikers: "If you want them to be Teamsters, you just tell them they're Teamsters!" (113). Such Anglo renaming would strike a deep chord in an audience celebrating its distinct cultural and racial identity and would point audience members to Chicano solidarity with other oppressed racial groups.[10]

At the acto's end, the military-agricultural complex still reigns. Butt Anglo may have second thoughts—who will buy poisoned food? why are we in Vietnam?—but they evaporate when the General reminds him that the military is buying his "scab lettuce" (119). Butt quickly understands—"It all comes back to me now"—and any moral qualms he might have had are swept aside when he remembers the money that his qualms might cost him (119). Butt and General Defense then sign a bill of sale, Butt with pleasure: "Con mucho gusto" (119).

Vividly illustrating the tragic effects of their alliance, the dead Hijo, drenched in blood, reenters, reminding the audience that the "war in Vietnam continues, asesinando familias inocentes de campesinos. Los Chicanos mueren en la guerra, y los rancheros se hacen ricos, selling their scab products to the Pentagon. The fight is here, Raza! En Aztlán" (119–20). His reference to Aztlán, the American Southwest as the Chicano homeland, suggests that Chicanos live in a militarily occupied, war-torn land similar to Vietnam (Harrop and Huerta 32 n. 11; Kourilsky 38). The Chicano and the Vietnamese families then rise and the acto closes with gestures of defiance: "*They all raise their fists in the air, in silence,*" in protest of the racial injustices of the military-agricultural complex, a gesture that recalls the black power salute (120). Signifying that the problems the acto confronts remain unresolved, this ending demands that the audience stand defiant, too.

## Taking a Stand in Louisiana

Both *Dos Caras* and *Vietnam Campesino* explore Chicano racial identity—the first distinguishing it from the greed of the Anglo's double self-image, the second identifying it with oppressed farmworkers worldwide. Created in truly communal ways, each acto assails industrial agriculture for its racially hierarchized system of labor, *Vietnam Campesino* casting a wider net to expose how federal manipulations of power serve that system. With the rise of Teatro Campesino, a full range of views about industrial agriculture is on the table: Norris's matter-of-factness, Cather and Glasgow's praise, Mitchell's uncritical acceptance, Valdez's outright protest. But more of protest is told in Ernest Gaines's *A Gathering of Old Men* (1983), a novel tackling issues of race and industrial farming in 1970s rural Louisiana. Just as Chicano theater gathered people to confront Anglo racial discrimination through communal storytelling, Gaines's novel brings together several old black men to stand against white racism through the stories they tell.

Frank W. Shelton argues that Gaines's novel "treats pastoralism in its full complexity" (14). Gaines appears to agree when he describes the novel: "And the whole damned thing is about remembering the past; they're living in the past. You're a southerner, I'm a southerner myself, and we all live in the goddamned past" (Gaudet and Wooton 103). Though this may be the case, a close reading reveals that the most important questions posed by the novel are not pastoral but georgic: What are the best types of work? What is the best way to work with nature? Is it natural that a complex rural community defined by shared obligations and responsibilities should be replaced by a complex urban industrial society defined by legal obligations and enforced responsibilities? Gaines's old men do not lament a lost past—they lament the fact that they allowed that past to unfold as it did.

Proclaiming solidarity with black southern farmworkers, Chavez named his headquarters—"*a barren, dusty plot by the city dump west of Delano*"—Forty Acres (Levy 220). The name alludes to the widespread belief among blacks following the Civil War that the U.S. government would divide up southern plantations among former slaves, each freedman receiving " 'forty acres and a mule' " (Royce 100; Fite, *Cotton Fields* 2). But southern farmland was never widely redistributed, and many blacks were soon drawn into a sharecropping system that perpetuated the injustices of the antebellum plantation. Sharecropping shoved most southern blacks into a perpetual cycle of poverty, a cycle enforced by Jim Crow laws that denied blacks basic civil rights (McGee and Boone 8, 19). In spite of these impediments, however, many African Americans did purchase farms, and by 1910 black farmers numbered 212,972 in the Blackbelt South, nearly double that of twenty years before (McGee and Boone 3). By 1920 black farmers in the United States totaled 926,000 and owned 15 million acres; but in 1978 black farmer numbers had fallen to just over 50,000, their landholdings to 3.2 million acres (Banks 1). These precipitous declines began immediately after World War I and have continued to the present to the point where scholars now predict the imminent disappearance of the black farmer (Schweninger 42; Browning 69).[11]

That racial discrimination and farm industrialization are big factors in the eradication of the black farmer is well documented. For example, the year before *A Gathering of Old Men* appeared, the U.S. Civil Rights Commission warned that "the effects of historical discrimination and structural inequities could result in the extinction of black farms in this country if immediate measures are not taken to counter the biases presently built into the system" (Browning 69). The commission claimed that "racism, a lack of institutional economic support, and possession of only marginal holdings" had contributed to black farmers' decline. More particularly, blacks were denied necessary farm loans because "financial institutions, including the Farmers Home Administration, have a reputation for discriminatory lending, which poses a real, as well as a psychological, barrier for blacks" (Browning 11, 63). This discrimination and inequity helps explain why "black-operated farms still remain, on average, abnormally small in terms of acreage and value of agricultural products sold" and why "the vast majority of black farmers do not have sufficient sales of farm products to survive on farm income alone" (Banks 17–18). After years of litigation, a class-action racial discrimination lawsuit brought by black farmers against the U.S. Department of Agriculture (USDA) ended in January 1999 with each farmer receiving $50,000 and federal debt forgiveness ("Judge Approves"). This came much too late for far too many.[12]

Farming in the South realized full industrialization only fairly recently: "southern agriculture changed marvelously little until well into the twentieth century" (Bartley 442; see Kirby 336–38). Not until the 1950s and 1960s did the South's "old rural system" totally collapse (Kirby 237). Agricultural historian Gilbert Fite notes that it was not until "the late 1930s to the 1960s [that] a revolution occurred on southern commercial farms. The fewer and larger farms became mechanized, capital intensive, and labor efficient" (*Cotton Fields* xii). One observer of southern farming claims that "'Moses and Hammurabi would have been at home with the tools and implements of the tenant farmer'"; Fite himself attributes "the slow progress of mechanization on southern farms" to the "large supply of cheap labor" in the South and "the lack of a machine to weed and pick cotton" (*Cotton Fields* 150).

As southern agriculture industrialized, undercapitalized black farmers working small farms were without means to compete. According to Fite, "An especially rapid exodus of blacks from southern farms occurred in the late 1930s. Some were 'tractored' off the land, but a much larger number left in hope of finding better economic opportunities in the towns and cities. They also hoped to escape discrimination" (156). The South's drift toward an increasingly industrialized agriculture left many black farmers at a disadvantage: "In the 1940s and 1950s the success of tractors, followed by mechanical harvesters, and finally by chemical weed control, led to the displacement of thousands of tenant farmers, most of them black" (Browning 38). Gaines's novel explores the intersection of the tractor and racial discrimination through characters that witnessed this era of change.

The loss of black-owned land and farms is significant beyond simple economics and numbers. Owning land has long symbolized freedom and independence among Americans, but especially among African Americans. Following the Civil War, "freedpeople were attracted to the possibility of landownership because of the independence and freedom it promised" (Royce 108). This attraction has continued down to the present: "studies have indicated that landownership by blacks tends to correlate highly with characteristics generally regarded as worthy of encouragement within the black community . . . landownership gives blacks a measure of independence, a sense of security, and the dignity and power which are crucial to the elevation of the social and economic status of the black community" (McGee and Boone xvii–xviii; see Browning 5–6).

Toni Morrison's *The Bluest Eye* (1970), an urban novel much concerned with earth, seeds, seasons, and a rural past, vividly portrays blacks' emphasis on landownership and that concern's connection to racism. Early in the novel, Claudia, one of Morrison's narrators, explains the "difference between being put *out* and being put out*doors*. . . . Outdoors was the end of something, an irrevocable, physical fact, defining and complementing our metaphysical condition" (18). Consciousness of this "outdoors," both concrete and abstract, "bred in us a hunger for property, for ownership. The firm possession of a yard, a porch, a grape arbor. . . . Renting blacks . . . look[ed] forward to the day of property" (18). Gary Grant, national president of the Black Farmers and Agriculturalists Association, makes plain what is at issue here: "A landless people is a helpless people."

Racially imbalanced landowning, "where whites control virtually all agricultural production and land development," puts more and more blacks outdoors and "can only serve to further diminish the stake of blacks in the social order and reinforce their skepticism regarding the concept of equality under the law," a skepticism mirrored in Gaines's plot, which centers on the old men's armed confrontation with white sheriff Mapes (Browning 8). Given all this, the stakes are high for the independent black American farmer, and they are even higher in the novel when one recalls that most of the old men have been denied even this status or have had it brutally stripped from them.

Gaines's old men take their stand against racist whites in the late 1970s in the quarters of a sugar plantation in fictional St. Raphael Parish, part of Louisiana's south-central sugar bowl. The quarters are lineal descendants of slave quarters on antebellum plantations, the place where plantation owners confined slaves in order "to observe, break, and control them" (Rowell 736). Despite such efforts, slaves developed a "culture, society, and world view that persisted, however modified, into the twentieth century . . . what we refer to as Black South folk culture originated in the quarters or from experiences with restrictions and forms of oppression not unlike those of the quarters" (Rowell 736). In the mid-twentieth century, the quarters were typically residences of sugar plantation wage laborers, like the novel's Charlie Biggs (Hoffsommer 11–12). The aging survivors of Marshall plantation represent the slowly deteriorating vestiges of a rich, African American rural folk culture; by the time in which the novel is set, most quarters, such as those at Marshall, had been almost completely depopulated, with only a scattering of old men and women left (Doyle 77).[13]

The racial situation in Louisiana is complicated by the presence of Cajuns, descendants of French exiles from eighteenth-century Acadia, who were looked down upon early by aristocratic French families already in Louisiana and later by Americans who moved into the region following the Louisiana Purchase. Placed in a social position between white landowners and blacks, Cajuns were Louisiana's "'poor whites'" (Doyle 78). Themselves "victims of caste and an unequal economic structure," Cajuns are blacks' chief competitors in places like Gaines's native Pointe Coupée Parish and "progress only at the expense of the Creoles and blacks" (Doyle 79).[14]

When plantations began breaking up in the twentieth century, owners rented mainly to Cajuns, who "used their racial connections with the owners to get the best lands" (Doyle 79). After acquiring prime property, Cajuns increased their agricultural production, rented even more land, bought the most up-to-date equipment, and kept abreast of the latest trends in the South's newly industrializing agriculture. In effect, they displaced poorer, sharecropping blacks, an eviction poignantly encapsulated in Tucker's story (Doyle 79). Whites' land use in the twentieth century has thus increasingly circumscribed southern blacks, both geographically and socially; as Chimley observes while fishing the St. Charles River, "The white people, they done bought up the river now, and you got nowhere to go but that one little spot. . . . Just ain't got nowhere else to go no more" (27).[15]

The bitterness lingering over the Cajun landgrab surfaces when Cherry invokes a title-by-toil claim to the plantation. He complains that Beau Boutan was leasing land, "The very same land we had worked, our people had worked, our people's people had worked since the time of slavery. Now Mr. Beau had it all. Or, I should say, he had it all up to about twelve o'clock that day" (43). Cherry's allusion to the Civil War Homestead Act's definition of land ownership—that he who works the land has the right to it—reflects how blacks have not been welcomed participants in that ideal even after the slavery question was decided. Just as slave farmworkers of antebellum days could not claim ownership of their labor, let alone land, postwar black farmworkers, though able theoretically to make such claims, cannot in fact, due to sharecropping, Jim Crow laws, and the rise of industrial farming. Cherry further implies that another civil, but this time racial, war, represented in Beau's high-noon death and the novel's final battle, will decide whether or not blacks will be landowners.

Beau Boutan's death, which sets the novel in motion, is the result of a dispute that makes sense only in the context of an industrial agriculture erected on the grounds of antebellum race relations. Boutan and Charlie Biggs, Beau's black employee, come to blows over whether or not Charlie has the right to his own labor: Beau attempts to deny Charlie his freedom to quit his job. In recounting his story to Sheriff Mapes, Charlie claims that he quit after Beau had "cussed" him and threatened to beat him, a treatment no man "half a hundred" ought to have to take (190). But when Charlie tries to walk off the job, Beau pursues him with a cane stalk; Charlie, for the first time in his life, resists, and the two battle until Beau goes down.

Fearful that he had killed Beau, Charlie "started running for the quarters" toward his godfather Mathu's house (190). Beau pursues on a tractor, jumps down, and comes at Charlie with a shotgun he kept "with him all the time" (191). Taking his "stand," Charlie holds Mathu's shotgun "steady as a rock" and shoots Beau in self-defense when Beau raises his weapon to fire (191). Had Charlie not defended his right to quit, he would have been no more than a slave; he would have remained "Big Charlie, nigger boy," someone to work, to abuse, someone Beau thinks he can "have a little fun with. . . . Was go'n hunt you like a rabbit, and shoot you when I got tired" (189, 191).

Fearing white retaliation, Charlie runs from the scene, leaving Mathu to shoulder the blame for something Charlie defines as having "started . . . forty-four, forty-five years ago," when he was a child in the 1930s, when industrialization was transforming southern farming (188). A survivor of that change and its accompanying racial conflict, Mathu waits to face the wrath of Beau's father, Fix Boutan, an avowed racist. But Candy Marshall intervenes and claims responsibility for the crime. At her behest, several old men gather at Mathu's with twelve-gauge shotguns to protect their neighbor and to face the sheriff and Fix's anticipated revenge.

As the novel's title suggests, the men's old age is important. Bearing witness to long suffering, they bring to the present, through their stories—their testimonies—the tragedies their community has suffered. In their seventies and eighties, these men reached their prime in the 1920s and 1930s, just after World War I—the beginning of "the long and tragic decline of black agriculture and land tenure in the South" (McGee and Boone 15). Thus the lives of these men describe the arc of African American farming's fall from its 1910 pinnacle to its near extinction today. Within them, this history still lives and cuts deep. For example, Mat, asked by his wife what is the matter with him, explodes: "you still don't know what's the matter with me? The years we done struggled in George Medlow's field, making him richer and richer and us getting poorer and poorer . . . Oliver. How they let him die in the hospital just 'cause he was black. No doctor to serve him, let him bleed to death, 'cause he was black. And you ask me what's the matter with me?" (37–38). His question might as easily be put to readers who assume that history has nothing to say to the present.

Their ages also remind readers that by 1978 black farmers were, at fifty-eight, on average seven years older than white farmers, an age difference heightened when one remembers Beau's age, which is around thirty (Banks 10, 60). Warning of the consequences of such an age differential, a 1986 USDA report notes that without the infusion of young blood into the black farming community "we can safely assume . . . [that] the ranks of black farmers will not stabilize and will almost surely continue to decline" (Banks 10; see Schweninger 55). The report proved prophetic: At the end of the twentieth century, only two hundred black farmers were under thirty-five (L. Turner). Gaines points to this desperate situation through the complete absence of young black farmers in the novel and the fact that most of the old men were never able to become independent farmers in an agricultural community dominated by white landowners like Jack Marshall who refused to confront their own histories.

Black and white farmers have responded to the 1999 settlement of the discrimination case against the USDA from widely separated perspectives, something that has not helped racial tensions in mixed-race farming communities. Protesting in Arkansas in the summer of 2000, black farmers asserted that the government was delaying payments to them; they understood this procrastination as a continuation of the injustices that brought on the lawsuit in the first place (Parker 3). Marching in Little Rock, they carried signs and chanted, "No farms, No food, No justice, No peace"

(Parker 3). At the same time, there was a backlash to the settlement among whites. Reacting to the lawsuit, one white farmer dismissed it out of hand: "It's just about getting money and yelling discrimination." More ominously, "threats of physical harm by whites and USDA employees" are among the allegations that black farmers brought before the USDA's Office of Civil Rights that summer (Parker 3). These responses to the suit may be rooted in blacks' and whites' radically different perceptions of the United States's history of racial discrimination.

The novel nods to racially divergent views of history by depicting disparate white and black perspectives of Marshall plantation's past. The difference in perspective is illustrated through the novel's structure—fifteen varied, mixed-racial points of view. That these perspectives exist is the subject of several characters' stories. The old men claim that whites, blinded by the agricultural progress represented by the tractor, do not comprehend its human costs, while they, the black men who have paid those costs, do understand, all too clearly. Watching the novel unfold, readers see what these black men see: that black farm communities have paid the price for farm "progress" (99).[16]

The tractor dominates the novel's opening pages. Readers first see it through the eyes of the black child Snookum, sent by Candy to round up the old men. Snookum sees the tractor as soon as he "hit the road": "I saw the tractor in front of Mathu's house. The motor was running, I could hear it, I could see the smoke" (5). Repetition points to its importance—Snookum looks "back at that tractor. The motor was still running" (6). Knowing that Charlie is as much Beau's tool as the tractor, Snookum questions Mathu as if Charlie's absence from its seat violates a law of nature, something as extraordinary as the sight of Beau's bloody body: "Where Charlie? . . . How come he ain't driving that tractor?" (6). The rest of the novel answers these questions: Charlie is not driving the tractor because he killed Beau.[17]

Whites not only own the tractor but they also control it. When Sheriff Mapes arrives, he orders his white deputy to cut its engine before he "turn[s] his attention toward the old men with the guns" (65). As if emphasizing the tractor's presence, Mapes repeats his order. Just as only he has the authority to silence the tractor, only he has the authority to question the men with impunity. The old men's attention shifts from the tractor and its roar to the sheriff and his violent interrogation, thus keeping before them—and us—the continuity of the racial oppression they feel (68–71).

After both Mapes and the tractor have been silenced, the air is clear for the old men to claim their voices.

Several of their stories couple racism and industrial agriculture. In recounting his reasons for his stand, Johnny Paul takes whites' refusal to see the human costs of "progress" as his theme. "Looking toward the tractor and the trailers of cane out there in the road," Johnny Paul hears Mapes say, "I see . . . I see" (88). "No, you don't see," Johnny Paul replies and asks Mapes what he does see when he looks at the quarters. Reflecting whites' inability to see from blacks' perspective, Mapes replies, "I see nothing but weeds" (88). What whites like Mapes do not see is the time "when they wasn't no weeds" (90), when a community existed before its destruction by a racist industrial agriculture, a community resurrected in the gathering in Mathu's yard: "We stuck together, shared what little we had, and loved and respected each other. . . . Where the people? . . . And where they used to stay, the weeds got it now, just waiting for the tractor to come plow it up" (91–92). Mapes's obtuseness recalls Valdez's actos when the Patroncito sees migrant life as pastoral and the General labels campesinos pests.

Where whites see a field grown to weeds, blacks see an identity lost, lost to deliberate attempts to erase that identity through the spread of industrial farming. Johnny Paul tells Mapes that he murdered Beau because "that tractor is getting closer and closer to that graveyard, and I was scared if I didn't do it, one day that tractor was go'n come in there and plow up them graves, getting rid of all proof that we ever was" (92). Emblematic of Johnny Paul's fear of blacks' erasure, the overgrown graveyard is where most of the men gather before going to the quarters to take their stand.

Sharing a history and a place, these neighbors assemble at a community plot, "the burial ground for black folks ever since the time of slavery," a ghost of the community that was: "They had all come from the same place, they had mixed together when they was alive, so what's the difference if they mixed together now?" (44). They gather to avenge that community and to preserve what remains: Johnny Paul knows that whites are "trying to get rid of all proof that black people ever farmed this land with plows and mules—like if they had nothing from the starten but motor machines" (92). His stand is a stay against this confusion: "Sure, one day they will get rid of the proof that we ever was, but they ain't go'n do it while I'm still here" (92).

Tucker's story proves that the triumph of industrial farming was not natural; powerful forces maliciously displaced preindustrial farming and those who practiced it. Tucker tells of his brother Silas, a man "they got rid of" before Candy was born, "the last black man round here trying to sharecrop on this place" (93). Many who challenged industrial agriculture were simply killed; Silas, for example, was "the last one to fight against that tractor out there" (93). Years ago, he and his mules beat Felix Boutan and his tractor in a pulling contest, but Silas won when he was supposed to lose: "How can flesh and blood and nigger win against white man and machine?" (96). Quickly beaten "down to the ground" for his victory, Silas dies, but not only at the hands of whites—Tucker himself helps, out of "fear" (97). Serving a racist industrial agriculture, the law took Boutan's side, saying that Silas had "cut in on the tractor, and he was the one who started the fight. That's law for a nigger" (97).

To all the men's stories, Mapes, representative of a white law that protects industrial agriculture, responds: "You ever heard of progress?" (99). In the novel's racial context, especially following his violent interrogation, Mapes's question is doubly ironic—his notions of social and technological progress differ sharply from that of the old men. Viewing many of the old men with disdain, he thinks of progress only in material terms and refuses to consider how it might wreck human lives; viewing all progress as inevitable, he cannot imagine humanity refusing a new technology. There was a horse-tractor debate before World War II, one the old men remember and Mapes cannot (Ellenberg 546). Though today we tend to believe that the horse's complete replacement was inevitable, the issue was always in doubt; at the very least, history need not have gone as it did. Witness the Amish.[18]

Though the old men take a stand against overt racism, they also stand against white paternalism, represented by Candy Marshall's unsolicited protection of Mathu and her constant references to blacks in the quarters as "my people . . . I won't let them harm my people. . . . I will protect my people. My daddy and all them before him did" (19). Such internalized and, perhaps, more insidious racism, was still common in the 1970s; in 1972, for instance, Murphy Foster, "son of a governor and U.S. senator and one of Louisiana's most prominent growers," told *Saturday Review* that " 'Niggers are the happiest creatures on God's green earth. . . . I've worked with these people for thirty years, and if they have a problem, they know to come to me' " (Schuck 37).

The old men refuse Candy's protection. When they want to talk privately together in Mathu's house, she will not let them unless she is present. And when Clatoo tells her that they do not want her, Rooster observes: "That stopped her. Nobody talked to Candy like that—black or white—and specially not black" (173). That her attitude alludes to pro-slavery arguments regarding blacks' inability to care for themselves is suggested when Mapes tells her, "And you want to keep them slaves the rest of their lives," to which Candy replies: "Nobody is a slave here . . . I'm protecting them like I've always protected them. Like my people have always protected them" (174). Candy's inherited protection of Mathu ends, as Lou explains to her: "When he pulled your hands off his arm and went into that room, he was setting both of you free" (184). Significantly, the novel ends not only with the black men victorious at the final gunfight, but also with Mathu refusing Candy's offer of a ride home: "he would go back with Clatoo and the rest of the people. . . . Candy waved goodbye to them" (214).

Pitting the white Luke Will, who wants to lynch Beau's killer, against the old black men who stand to stop him, the novel's shootout links racism and industrial farming. While Mapes lies helpless in Mathu's front yard, Will's gang shields itself with the tractor (197). The old men who surround the tractor fire at it as much as at those it conceals: "Mat, Jacob, Ding, Bing, fire at that tractor" (199). In the battle's climactic confrontation, Charlie, a black man who has been unable, until that day, to define himself as a man (186), confronts Luke Will, a white man who has kept "things running smoothly in the parish" through intimidation of blacks (159). Speaking for all the black men, Charlie tells Lou Dimes, "I'm standing up to Luke Will" (208). Firing, Charlie "headed straight toward that tractor," gets shot, staggers, but again goes "toward that tractor," finally falling (209). Appropriately, and once more tying industrial farming and racism, Will, the lynch gang's ringleader, dies propped up "against one of the tractor wheels" (209).

Though Charlie dies too, he wins a moral victory; the others "leaned over and touched him, hoping that some of that stuff he had found back there in the swamps might rub off" (209–10). But what they hope for they have already achieved—as Charlie tells Dirty Red, "All of y'all seen it" (208). They have all finally had the courage to stand up against abuses heaped upon them, as much by telling their stories to Mapes—and us—as they do by firing their shotguns at Will and the tractor.

Representing a tension between rural and urban perspectives, the novel suggests that the old men's challenge to industrial farming never gets a full hearing. Though the old men tell their stories to each other, to readers, and to the court, an urban media redefines events at the plantation "as something astonishing but not serious" (212). And though readers know full well the violent racial tensions in rural St. Raphael Parish, urban America knows only that the Louisiana State University (LSU) football team has a racially balanced backfield. Two other urban media filters compete with the old men's stories: four chapters are told through the perspective of Lou Dimes, a reporter for the paper in Baton Rouge, and events in two other chapters are described through Sully, Gil's teammate at LSU. The white Sully, who accompanies Gil Boutan to Marshall and to the Boutan homestead, is known among blacks as "T.V.," the main medium through which the world knows St. Raphael Parish.[19]

The possibility of racial cooperation, the novel's final message, is played out not in a cane field but on a football field in a nationally televised game. The LSU team features Gil Boutan at fullback and Calvin Harrison at halfback. Together, they are Salt and Pepper, on the road to becoming all-Americans, "the first time this had ever happened, black and white in the same backfield—and in the Deep South, besides" (112). Not incidentally, the offense is led by quarterback "Sugar" Washington, completing the novel's racial triad—white, Cajun, and black (112). But unlike in the cane fields, on the football field the races work together; Gil and Cal "depended on each other the way one hand must depend on the other swinging a baseball bat" (115). Sully, LSU's third-string quarterback, notes that if LSU beats Ole Miss "nobody else could stop us, and we would host the Sugar Bowl game on New Year's Day," a more watched nationally televised game—thus more widely showcasing racial cooperation (112). If racial harmony can win the Sugar Bowl, maybe it can win Louisiana's sugar bowl, too.

After refusing to avenge his brother's death, Gil must make another moral choice—whether or not to play against the Ole Miss Rebels, who will win if they "stop one or the other" of LSU's backfield (112). Russ, the police officer charged with keeping Fix and his cronies from riding on Marshall, tells Gil to play because "Luke Will and his kind don't want to see you and Pepper in that backfield tomorrow. He doesn't ever want to see you and Pepper together" (150). Thinking that he has hurt his father enough already, Gil objects, but Russ, tying the game to race and farming, replies: "Sometimes you got to hurt something to help something. Sometimes you have to plow under one thing in order for something else to grow. . . . You can help this country tomorrow" (151). Looking to the future, Gil chooses to play, denying Ole Miss's Rebels a win and Luke Will the satisfaction of destroying a sign of racial accord.

Underscoring the dawn of this new day in racial cooperation, images of communal harmony dominate the novel's close. For example, Gil is back in the Boutan family; he sits with his father at the trial (212). All the defendants, black and white, appear in court in similar physical conditions and dress: "Everybody had something wrong with him—scratches, bruises, cuts, gashes. . . . Everybody was either limping, his arm in a sling, or there was a bandage round his head or some other part of his body. . . . They had all taken baths, and wore their best clothes" (211). Laughter points to the conciliatory mood in the courtroom: the judge is "always happy . . . and has a great sense of humor" (212); hearing the old men's nicknames "bring[s] the court to laughing" (212); and "everyone in the courtroom started laughing" after Mapes testified that he was "sitting on [his] ass in the middle of the walk" during the gunfight (213). All defendants, "black and white alike," are given the same sentence, five years of probation, itself an act that puts these men back into their community to get along (213).

Placing everyone on probation fittingly ends this novel's study of race relations and industrial farming, for, in effect, the nation itself is on probation regarding these issues. Gaines and Valdez pose tough questions: How do we devise a sensible system of food production that will take into account the human costs of material progress? How should our methods of food production and distribution reflect our national political ideals of equality and justice for all? How will farm labor be treated—how will all labor be treated, for that matter—as sources of food become more and more industrialized and vertically integrated? How does our conception of agriculture affect how we perceive and deal with the rest of the world? So far, we have not addressed these questions in earnest because for far too many of us the struggle to feed ourselves amounts to choosing the shortest checkout line at the grocery store.

Literary works like Valdez's actos and Gaines's *A Gathering of Old Men* remind us that these questions haunt the edges of our consciousness. Giving shape and substance to moral decision-making, they invite us in the act of reading them to make moral choices along with their characters—choices that increase our capacity for making such decisions in daily life. As communal acts, novels and plays point us beyond our lives to sympathy with the lives of others. If we measure ourselves by those we know, reflecting on literature and its cast of thousands can refine our communal standards of judgment. If nothing else, literary works make these standards visible and open them to debate. The more complete and fair are our standards, the more likely we will live responsibly in—and with—the world. Issues of restraint and responsibility mark the work of Jane Smiley and Wendell Berry, two writers most recently wrestling with the question of what constitutes the most humane agriculture.

**Of special note was the idea that the agribusiness industry must do what is right for the farmer even though it may lose a customer.—Everett Newswanger, *Lancaster Farming*, 24 December 1994**

CHAPTER FIVE

# From *A Thousand Acres* (1991) to "The Farm" (1998)

In form and content, Everett Newswanger's Christmas Eve article concisely sums up the essential characteristics of contemporary agriculture's thinking and practice. Newswanger reports on a Lancaster County (Pa.) Cooperative Extension Service meeting held to discuss the future of Lancaster agriculture. Because he is the paper's managing editor, one can safely assume that his work reflects the perspective of *Lancaster Farming*, a regional weekly farm publication covering mainly eastern and south-central Pennsylvania. A multisection paper, *Lancaster Farming* carries market reports, agribusiness advertisements and classifieds, extension agent columns, and articles and photographs about farm-related events. Like much agricultural journalism, Newswanger's 309-word text, "Agribusiness Professionals, Farmers Gather to Discuss Future," presents itself as a news article, complete with dateline and byline; when read closely, however, the piece emerges as a defense of industrial farming.

The text privileges agribusiness "persons" at the expense of farmers. Significantly, agribusiness people far outnumber farmers at a meeting about the future of Lancaster farming: "A group of 60 agribusiness persons and a few farmers convened early Tuesday morning." Agribusiness counts because its representatives are counted, while farmers are discounted as "few." Though the agribusiness "persons" are anonymous—Newswanger does not even name the businesses they represent—they are characterized as "Professionals," according to the text's title, and "industry leaders," according to its concluding paragraph. Readers know nothing about the farmers who quickly drop from sight.

The meeting's stated purpose, identifying "internal and external problems facing [Lancaster] county agriculture," serves industrial farming. Functioning as a problem-defining cooperative, participants divided into small groups and then "forwarded specific ideas" to the group at large. All agreed that at least nineteen problems confronted the county's farming. Newswanger simply lists all nineteen because he is sure that his audience will know what makes, for instance, "animal rights" a problem. The phrase itself says little, but it vexes industrial farming because "the issue of farm animal rights is particularly pertinent to any discussion about technology in farming" (Wojcik 79). This makes it a potentially explosive public relations issue for farmers raising hogs in huge farrowing systems, chickens in assembly-line poultry houses, or veal in slatted-floor pens. Of course, such operations are capital-intensive, high-input, good-for-agribusiness enterprises.

Each identified problem has a strong tie to at least one of industrial farming's characteristics: monocultures, heavy capital investment, strict management-labor hierarchies, and economic measures of success. For example, an agriculture characterized by "resistance to change" will not purchase agribusiness inputs fast enough to expand agribusiness bottom lines. An agriculture saddled with "excessive regulations" has been restrained from using certain herbicides, pesticides, or chemical fertilizers to maintain single croppings. An agriculture that hierarchizes labor and management will always see labor as a problem, be hurt by a "lack of management ability," and lament a "lack of record keeping." And an agriculture defined solely by profit will worry over "consumer education and product promotion," "high fixed costs," and an "unfavorable tax climate." When the article names as problems "wrong measure of success" and "farming as a way of life," it assumes that the only way to measure successful farming is in dollars and cents. Imagining "farming as a way of life, not a business" might lead farmers to think long-term about the consequences of their decisions, beyond end-of-the-year cost accounting, toward the costs those decisions might incur for future generations, thus leading them to husband resources and to reduce reliance on agribusiness wares.

Newswanger has assented to agribusiness values without knowing that he has. When the first-named problem—"agriculture's resistance to change"—resurfaces, he speaks as a meeting participant instead of its recorder; his article becomes an editorial. According to Newswanger, meeting participants agreed that nonfarm industry sets the pace for agricultural change: "Comparing agriculture to non-agriculture, we know other industries have changed. And we need to help farmers make the needed changes." Who are "we"? Farmers are not included, apparently, because the piece defines them as "they." Most likely, "we" includes Newswanger and the agribusinesses represented in *Lancaster Farming*'s advertisements, a fact implied by his next sentence: "Of special note was the idea that the agribusiness industry must do what is right for the farmer even though it may lose a customer." The arrogance is too obvious: the "agribusiness industry" will decide what is right for the individual farmer, even if it kills him or her. Given this, one might ask who is to "stay in business," farmers or agribusinesses? In closing, the article notes that County Agent Shirk shows "industry leaders" a new, land grant–created input: "the Dairy-MAP education program from Penn State." One wonders where the meeting's "few farmers" disappeared to.[1]

Defining Lancaster farming as industrial is ironic in a county whose most famous farmers resist conventional practices in favor of preindustrial methods. Because Amish farmers practice technological restraint, those at the meeting do not profit from them. Amish emphasis on thoughtful change, horse-drawn field equipment, and community and family labor makes them a poor topic of discussion in the thinking of an agriculture whose profits end up mainly in agribusiness coffers and not in farmers' pockets.

Contrary to the perception of most people, the Amish are not leftovers of antebellum nineteenth-century agriculture. For example, the Amish do use tractors—but in barnyards, not in fields (Kraybill 175). The Amish argue that horses are more profitable than tractors and that limiting tractors to the barnyard maintains collective, community fieldwork (Lodgson 131; Kraybill 175–76). Change among the Amish occurs, of course, but not at the expense of the community's health. And they are aware of modern technology. Their technical innovations in ground-driven field machinery have had "input from some of the most sophisticated mechanical design engineers in the nation" (Lodgson 135). The Amish simply ask each other what is best for their communities in the long run, rather than what is best for the individual in the short run.

What the extension meeting did not consider a problem is as important as what it did; its omissions reinforce the fact that industrial agriculture ruled its thinking. Among the most glaring silences concern the environmental impact of Lancaster farming practices, the social health of the county's rural community, keeping farmers on the land, and maintaining a diversity of crops and animals. All are urgent issues in a world where Brazilian rain forests burn to make way for cattle ranches, where the fourth largest body of inland water has been siphoned off to irrigate cotton (Mathews), where a multinational corporation polices fields to keep farmers from "unauthorized seed-saving" ("U.S. Patent"), and where debates flare over the existence, let alone the future, of the American family farm (see Comstock).

## On the Grid with *A Thousand Acres*

The farm future envisioned in Newswanger's article is given a history in Jane Smiley's *A Thousand Acres* (1991). As contemporary as *A Thousand Acres* is, its first paragraph returns readers to the birth of the nation, when American capitalism was blossoming, when the industrial revolution was getting its full head of steam, when American farmers were pushing west in search of new lands. At this historical moment, Thomas Jefferson was not only promoting a national agrarian economy; he was advocating a new land survey system. Perhaps more than any other single factor, the intersection of this survey system with Jefferson's agrarianism set into motion the forces that have brought American farming to its present industrialized form.

Jefferson chaired the committee that created the new land survey system. His grid survey facilitated the distribution of land in an orderly, democratic manner that was supposed to realize his dream of a nation of independent husbandmen. The 1785 Land Ordinance, the act that opened up to settlement today's Midwest, called for surveying "townships six statute miles square divided into thirty-six lots of 640 acres each; lines of longitude and latitude determined the alignment lines" (Stilgoe 103). Though the specifics of Jefferson's committee report were not passed by the Confederation Congress, its ordering of the wilderness prevailed. It was upon—and in part because of—this grid that railroads were built, land grant universities were founded, and farms were established.

At the survey grid's heart lay a prophetic irony: the prototype survey for settling an agricultural West was "the *urban* grid of Philadelphia" (Stilgoe 99). Abstracting wildness out of existence, the grid replaced it with an urban checkerboard whose "psychological impact . . . cannot be overstated . . . the survey gave the impression of orderly, controlled, legal land development under the umbrella of government regulation and management" (Opie xx). After 1785, local tradition and simple geography no longer determined boundaries; land was defined geometrically. Ultimately, the grid system determined "the spatial organization of two-thirds of the present United States" (Stilgoe 99).

Centrally controlled from Washington, D.C., the grid embodied from birth an urban rationality that made rural areas colonies of cities and ensured that eighteenth-century subsistence farmers would become twenty-first-century contract laborers for agribusiness giants. As a network of lines criss-crossing the continent, the grid ignored natural land forms, hid local complexity with a comprehensive simplicity, facilitated the easy exchange of farmland, and created an attitude toward land that made it an abstraction rather than a concrete reality. A good example of the grid's order is a county map of Iowa, the setting of *A Thousand Acres*. A good illustration of the grid's effects on farming and farmers is the novel itself. A defining motif in *A Thousand Acres*, the grid is the exact measure of industrial agriculture's unrestrained desire to monopolize nature.[2]

The opening of *A Thousand Acres* nods to the grid by orienting readers to the Cook farm through descriptions of roads, which run along grid section lines: "At sixty miles per hour, you could pass our farm in a minute, on County Road 686, which ran due north into the T intersection at Cabot Street Road" (3).[3] The grid's right-angle ordering of the land's surface corresponds to a similar network below; a system of "magic lines of tile" drains this once marshy part of Iowa to enable row crop farming to occur—a system that narrator Ginny Cook describes as "a floor beneath the topsoil, checkered aqua and yellow" (15). The grid hides industrial complexity with comprehensive simplicity; tile and grid lines rearrange a once wild nature into an industrial landscape, an artificiality that only appears natural to the novel's nonfarm reader.

Speeding along the grid line, the reader enters the novel as a spectator—"you"—who sees only a farm landscape, not the land; the farm scenery, not the lives within it. That this is the customary position of someone passing a farm, or opening a farm novel, is suggested by the fact that the road soon turns into "Zebulon County Scenic Highway," a thoroughfare for sightseers whose connections with nature are merely visual (3). Hidden here beneath the land's surface flows the Zebulon River, which "made its pretty course a valley below the level of the surrounding farmlands" (4). Most likely an urban dweller, the reader may be lulled into a sense of nostalgia for this beautiful scene. But the novel brings the reader to the Cook farm, and there the reader comes to know the place and its inhabitants, quickly learning that the pastoral farms he or she speeds by are not as idyllic as they appear.

*A Thousand Acres* asks the reader to peer beneath the grid's surface to uncover the Cook farm's hidden corruption. By reading the lay of the land more closely, the reader understands the brutality behind the farm's reliance on monocultures of corn and beans, its dependence on pesticides and chemical fertilizers, and its specialization in hogs. But to see all this, the reader needs a vantage point from which to survey Larry Cook and his "six hundred forty acres, a whole section . . . flat and fertile . . . exposed" (4). The view from the T intersection is the "middle distance" perspective that Ginny and the reader use to watch the novel's events unfold (20). Returned to that middle ground at the novel's end, the reader is left alone with the grid: "at the intersection of County 686 and Cabot Street Road now, you see that the fields make no room for houses or barnyards or people. No lives are lived any more within the horizon of your gaze" (368).

Though its template is Shakespeare's *King Lear*, the novel is not a simple one-to-one correspondence to the play. In an interview, Smiley points out that she uses farming as a metaphor to represent the entwined abuses of women and nature: "Farming was something I'd always been interested in. So it wasn't as though I had the thought about using *King Lear* and found a place to put it. These two paths converged as a result of realizing that nature and women were so closely connected as exploitable objects in the mind of our culture" (37). As much as it is a feminist response to *Lear*, *A Thousand Acres* also answers Thomas Jefferson's famous declaration in *Notes on the State of Virginia* (1787): "Those who labour in the earth are the chosen people of God . . . whose breasts he has made his peculiar deposit for substantial and genuine virtue. . . . Corruption of morals in the mass of cultivators is a phaenomenon of which no age nor nation has furnished an example" (217). Smiley responds with a woman's voice, a voice generally unheard when Jefferson was arguing for a national agrarian economy.[4]

Her narrative challenges the idea that American agricultural history is a linear pattern of male achievement. First-person narrator Ginny (Virginia) Cook answers this version of history with her story. Whereas her husband, Ty, believes in an "ordered, hardworking world" rooted in Jeffersonian agrarianism, Ginny sees history as men "taking what you want because you want it, then making up something that justifies what you did" (370, 342). Where Ty sees a "grand history," Ginny sees "blows" (342). The male story she revises is dominated by her father, Larry Cook, a self-absorbed farmer who works his land and family without restraint, exploiting neighbors for land and his daughters for sex—all leading to the poisoning and, ultimately, the destruction of family and farm. Her narrative describes industrial farming as a mode of perception that suppresses wild nature, silences women, erases margins, and knows no limits—all the while projecting itself as perfectly natural and " 'right' " (343).

The novel describes Ginny's awakening to the fact that she has been living within a male vision of reality. For example, as a child Ginny was taught that she was a "beneficiary" of the fact that "it took John and Sam and, at the end, my father . . . to lay the tile lines" that created the Cook farm's wealth (15). But moving off the farm gives her a vantage point from which she mocks "the proud progress from Grandpa Davis to Grandpa Cook to Daddy. When 'we' bought the first tractor in the county, when 'we' built the big house, when 'we' had the crops sprayed from the air, when 'we' got a car, when 'we' drained Mel's Corner" (342). In recognizing her own exclusion from this "we," she sees the pattern in her family's history: only men defined the farm. To impose rigid mechanisms of control, men drained the land of its original fluidity and silenced women like her mother and her grandmother Edith—the latter's "reputed silence," for instance, "due to fear" of the men who surrounded her (132–33). Even the streets of Mason City graph the exploitation of women and land: a U.S. Department of Agriculture office and a whorehouse stand beside a restaurant (59). The diner shares a public space with a building where men exchange money for sex and with a federal agriculture department that teaches the mining of nature—all in a city whose name alludes to a secret international fraternity.

Though the novel is at pains to prevent it, early reviewers sped by the Cook farm, holding firmly to an urban passerby's perspective of farm as pastoral site. In the *New York Times Book Review* Ron Carlson claims that "Ms. Smiley's portrait of the American farm is so vivid and immediate—the way farmers walk, what the corn looks like, the buzz of conversation at the community dinners—that it causes a kind of stunning nostalgia." In her *Library Journal* review Ann H. Fisher notes: "Smiley lovingly creates an idyllic world of family farm life in Iowa in 1979; the neat yard, freshly painted house, clean clothes on the line, and fertile, well-tended fields." Confronted by a novel about attempted murder, suicide, adultery, cancer, miscarriages, and incest, reviewers still see only "well-managed acres" (A. H. Fisher)—skipping the fact that those acres have been unjustly acquired (135), chemically maintained (187), and tied to Larry's financial and sexual lust.

Reviewers' naive nostalgia resonates with cruel irony when alluded to in the novel; the Kansas man hired to construct a new hog confinement building tells Ty, "If this had been my dad's place, I never would have left. This looks like paradise to me" (178). And for farm men, it is. But by this point readers have learned about spousal abuse (31), schemes to displace neighbors (134–35), and nitrates poisoning well water (164)—all in preparation for Ginny learning that she is a victim of Larry's incest (189). Remembering the incest reorients her to a "new life, yet another new life" in which "anything could happen" (229, 225).

Lording over his paradise is Larry Cook. Ginny's childhood memories of her father imagine a man rivaling God in her universe; when other children claimed that their fathers were farmers, she did not believe them: "To really believe that others even existed in either category [i.e., farmer and father] was to break the First Commandment" (19). Larry's godlike qualities fit industrial agriculture's faith in a bigger-is-better philosophy, reflected in the fact that the Cooks' farm was "the biggest farm farmed by the biggest farmer. That fit, or maybe formed, [Ginny's] own sense of the right order of things. . . . He was never dwarfed by the landscape—the fields, the buildings, the white pine windbreak were as much [Larry] as if he had grown them and shed them like a husk" (20). In short, Larry is how he farms.

Larry Cook's bigger-is-better farming attitude is made possible by the grid's theoretically infinite expanse. Anticipating the unlimited growth ideal of industrial farming, the grid "had no final boundary line. It measured and described public lands, known or unknown or imagined" (Opie

3). Cook and his neighbors are clearly more interested in extending themselves from grid line to grid line to expand their bottom lines than they are in maintaining a community of neighbors and friends. The Cook-Clark rivalry over neighbor Cal Ericson's land is one example; another is the shady deal Larry's father makes to ace the Stanleys out of acquiring a grid angle, Mel's Corner (134–35).

Larry's perception of the world is grounded in an inordinate desire that is most visibly expressed through his expansion of his industrial farm. He has a "lust for every new method designed to swell productivity" (45). Long before he raped her, Ginny understood that her father defines himself as a man who knows no bounds: "What is a farmer? A farmer is a man who feeds the world. What is a farmer's first duty? To grow more food. What is a farmer's second duty? To buy more land" (45). In a 1957 *Wallace's Farmer* story entitled "Will the Farmer's Greatest Machine Soon Be the Airplane?" Larry is quoted as claiming: " 'There isn't any room for the old methods any more. Farmers who embrace the new methods will prosper, but those that don't are already stumbling around' " (45). As he says this, he looks across the road to the unruly farm of the Ericsons, who keep cows " 'because they like them' " (44). Larry's absorption of the Ericson property redeems it; his neighbors' loss is right because "a poor-looking farm diagrams the farmer's personal failures" (199). Where Cal Ericson farmed to "enjoy himself as much as possible," Larry farmed with "a sort of all-encompassing thrift that blossomed . . . in the purchase of more land" (44–45).

Because the grid remains the same from point to point, from moment to moment, it is outside history—and responsibility. In their relation to the grid, humans merely come and go as interchangeable marks on county assessor plats. At home on the grid, Larry believes that "history starts fresh every day, every minute, that time itself begins with the feelings he's having right now" (216). As the grid does for landforms, so "incorporation for Cook represents a complete melding of identities rather than the maintenance of familial distinctions or paternal preferences" (Leslie 36). Larry's perspective denies accountability; seeing everything as an extension of himself, he cannot distinguish wives and daughters or cattle and humans (10). Such thinking, of course, justifies whatever one is doing at the moment; ignoring history frees one to farm and live like there's no tomorrow.

The erasure of native bodies of water in the novel is emblematic of the grid's uniform conception of land, regardless of its local peculiarities.

The Cook farm, like most of Zebulon County, is naturally marshland, but farmers have drained so many ponds and lakes that communities now build public swimming pools (87). The history of the county's agriculture has been one of vigilantly suppressing the land's natural inclinations. When the Davises purchased their land in 1890, it "was under two feet of water part of the year and another quarter of it was spongy" (14). Like many pioneering settlers, Ginny's great-grandparents bought land at a distance, "sight unseen"; they were oriented to their properties before taking legal possession by locating coordinates on a land office map (14). The Davis and Cook men spent "a generation, twenty-five years," laying enough tile to drain it and fashion a new landscape: "There was no way to tell by looking that the land beneath my childish feet wasn't the primeval mold I read about at school, but it was new, created by magic lines of tile. . . . However much these acres looked like a gift of nature, or of God, they were not" (15).

Ginny comes to know that the new landscape was not fashioned for women: her earliest memory is sitting on a drainage grate, the "filmiest net of the modern world," feeling that her life had "vanished into the black well of time" (47). This farm landscape kills off the women who live within it: Jess Clark's mother (53) and Rose die of cancer, most likely caused by nitrates in the area's well water (164–65, 370). Ginny's mother dies of an "illness" (91); after her death she "disappeared"—all material reminders of her existence were stripped from the Cook home (227). Larry's sisters die in an influenza epidemic that Larry survives (132); his mother dies young, at forty-three (132).

Larry's draining and filling of Mel's pond is a metaphor for his raping of his daughters. In each case he bulldozes distinctions: just as the difference between land and water disappears, so does the difference between wife and children; all are his "possession[s]" (191). In the early sixties, when "fences [were] coming down," when Ginny was in her teens and Larry was raping her, Mel's Corner was redefined: "bulldozers were a sight [she] had glanced at from the window of [her] bedroom" (206). Years later, after Rose tells her about the incest, Ginny revisits Mel's Corner, on Independence Day, "looking for signs of the old pond, but . . . the rows of corn marched straight across black soil as uniform as asphalt" (205). Lost and searching for her preincest life, she cannot find the pond: "How many times had I walked this way . . . knowing precisely where I was going and what pleasures were to come? But in the leafy rows of corn I did not find even the telltale dampness of an old pothole to orient myself" (206). What had been a rich landscape is now a monoculture whose "manlike" corn and "mechanical uniformity" confuse her (152). What had been her confident childhood is erased by her father's regular violations of their father-daughter relationship.

Though alternatives exist to the uniformity of the industrial farming portrayed in the novel, they are short-lived. Jess Clark, Ginny's lover, hopes to establish an organic farm. But his father publicly humiliates him, making organic farming sound like a conspiracy to seize the Clark farm: "You got your eye on my place. . . . Harold, you ought to do this! You ought to do that! Green manure! Ridge till cultivation! Goddamn alfalfa!" (219). Though organic farming's more careful relationship with the land never finds a place in this Iowa, wild nature does crack the grid's surface: along the Scenic River, at the quarry, and at the farms' dump "behind a wild rose thicket" (122). At the farms' margins, the dump is where native big bluestem still grows—through "the metal grid of the bedstead" (124). There Jess and Ginny secretly meet, and she wonders if "that wasn't the right way to look at things after all, standing in the dump, smelling the wild roses" (128). Almost immediately she and Jess Clark kiss for the first time; later, they make love in the bed of the old pickup there (162). But Jess, the sisters' alternative to their husbands and the novel's spokesman for alternative farming, betrays Ginny for Rose only to abandon her and their organic farm.

## The World Series of Monopoly

Like Iowa agriculture, Monopoly is played on an urban grid, across which players move to buy properties on which they must make capital improvements to survive. The goal of the game, of course, is to take everyone else's money and become ruler of the board; the only restraint on any player is economic. Neither the game's grid nor the survey grid acknowledges any community outside a ruthlessly competitive urban paradigm. Juxtaposed to Jess Clark's wishful thinking about practicing a diverse organic farming, the Monopoly tournament in *A Thousand Acres* is played in contrasting city and farm styles. Pete, characterized earlier as wanting "to make a killing" in farming (31), is "an aggressive Monopoly strategist," whereas Rose, Ty, and Ginny play "like farmers, looking for the pitfalls" (78, 77). When the tournament ends, the aggressive Pete is winning the last game. Though urban monopoly means the quick score, the double-or-nothing attitude, and though rural monopoly play might be more conservative, both seek the same goal: winning by attrition.

To illustrate the success of urban monopoly play, *A Thousand Acres* sweeps characters from the landscape. Beginning with a countryside divided among three farms, the novel ends with the land owned by a single urban agribusiness giant, the Chicago-based Heartland Corporation, which specializes in hogs raised in confinement systems. These centrally controlled food systems also absorb the novel's main characters. After selling the farm, Ty goes to Texas to get "a job" in the "big corporate hog operations" there (339). Rose's daughter Linda, a prebusiness major in college, is "especially interested in vertical food conglomerates, and may go to work for General Foods" (369). Living without even the "visceral experience of the natural world" that she had enjoyed as a child, Ginny works as a waitress at a Perkins restaurant in St. Paul (Mathieson 141).

Moving away from the farm may help Ginny to reorient herself, but she remains captive of grid and farm. The standpoint she achieves is the acute awareness that her body is an intersection of grid lines (Slicer 69). Enjoying the "blessing of urban routine," she lives her "afterlife" in a garden apartment along a Minnesota interstate (336, 334). And she is literally made up of elements of an industrial farm: "Lodged in my every cell, along with the DNA, are molecules of topsoil and atrazine . . . and also molecules of memory . . . each particle weighs . . . perhaps as much as the print weighs in other sorts of histories" (369). As a waitress, she merely moves from one end of the food delivery system to the other. Instead of being a farmwife slavishly preparing and serving food for an industrial farm family, she ends up at a roadside restaurant waiting on truckers (337).

In its movement from community pig roast to absentee owner hog raising, *A Thousand Acres* describes industrial agriculture hurtling to its logical conclusion. In the end, only a factory in the field remains on the grid where several farms used to be; all traces of community have been erased: family members are scattered or dead, homes removed. Ginny's house, for example, is "taken down to make room for an expansion of the hog buildings to give them a five-thousand-sow capacity" (368). One may recall the leveling of Broderson's place in *The Octopus* to make way for the larger demands of the New Agriculture.

Understanding that Smiley nods to the American survey grid system in *A Thousand Acres* leads us to see her novel not simply as a brilliant reworking of *King Lear*, but as a significant comment on the history of American agriculture. Understanding that this history was determined by an Enlightenment invention that situated a single male story on the American landscape, we might want to extend the novel's epigraph from Meridel Le Sueur to read: "The body repeats the landscape" which repeats the grid. "They are the source of each other and create each other."

## Butz to Berry

In the satirical novel she published four years after *A Thousand Acres*, Jane Smiley makes clear that this single male story is not natural; it is taught. *Moo* (1995) ties the bigger-is-better attitude of industrial farmers like Larry Cook to the land grant university system. At *Moo*'s center is a hog named Earl Butz, who is raised by an ag professor simply to find out how big he will grow before he dies (6).[5] Set at a midwestern university nicknamed Moo U, the novel depicts the nation's well-documented incestuous relationship between land grant universities and agribusiness industry (Hightower and DeMarco, 146–49). Moo U professors labor to create closed systems that will forever produce seamlessly perfect products like the Cooks' hog confinement system.[6] For example, an animal science professor dreams of "beautiful black and white Holsteins in a green pasture, all marked the same, all turning their heads, all mooing, all switching their tails, all in unison, a clone herd, the perfect herd of perfect cows" (57). His latest pitch to ag corporations is a plan to create "unending lactation" in cows by artificially inducing false pregnancy (99). Texas billionaire Arlen Martin is interested in funding the project in exchange for patents it will likely yield (131). Martin, head of TransnationalAmerica Corporation, also approaches the financially strapped—and taxpayer-supported—Moo U with a plan to make the university his company's research and development (R and D) arm. As Martin points out to Moo U's president, "Why should I hire R and D people just to read what your R and D people already know?" (76). Asked of one corporate head by another, his proposal has a perfectly logical answer: "Of course" (76).

Farmers like Larry Cook became paper millionaires because of the farm policies of Nixon agriculture secretary Earl Butz. Following the Soviet entry into the world grain market in 1972, Butz urged farmers to "plant from fencerow to fencerow" to meet the new demand (Fite, *American Farmers* 202; Strange 18). Grain prices shot up: soybeans by 52 percent; corn, by 92 percent; wheat, by 132 percent (Strange 18). With interest rates low and inflation high, government officials, university experts, and major farm organizations urged farmers to buy more land to plant ever more crops. Seeing prices high and heeding the experts, American grain farmers went deeply into debt to expand their acreage. As a result, prices for farmland skyrocketed: from 1970 to 1980 the value of American farmland quadrupled—from $176 billion to $715 billion. But between 1970 and 1981 farm debt borrowed against the value of land jumped from $29 billion to $96 billion (Strange 19, 22). When farmers complained about their debt or about being priced out of the market, Butz famously told them, "Get big or get out," a dictum taped over the pen of *Moo*'s Earl (4). In 1979, after the Federal Reserve tightened the nation's money supply, interest rates rose, inflation fell, and the farmland boom went bust. By the mid-eighties farms were rapidly foreclosing, farm-related businesses were going bankrupt, rural banks were shutting their doors, and rural towns were teetering on the edge of collapse. The 1980s Farm Crisis was on, and agriculture was plunged into its worst depression since the 1930s.

Butz and present-day defenders of industrial agriculture have argued that the U.S. food system is well ordered, fail-safe, and inexpensive, that the cornucopia it provides is abundant and diverse, and that problems in the delivery system are merely glitches that any large-scale operation will run into from time to time. They point out that Americans no longer need to grow their own food and that they should take solace in the fact that there are families down on the farm willing to take on the burden of providing it. The "agriculture industry" runs smoothly and will continue to do so as long as farms are freed to get bigger, more "efficient," more technologically up-to-date—an outlook overlaid with nostalgia, like that in Archer-Daniels-Midland (ADM) "supermarket to the world" commercials aired mainly during Sunday morning political talk shows.[7]

## An Alternative Paradigm

Recent critics of industrial farming counter that behind the ordered veneer lurk significant problems. Bigger farms are not necessarily better or more efficient (see Strange 78–103). Large farms, they claim, mine the soil and poison rivers and lakes; glitches in the food system are symptomatic of how fragile that system really is. Critics point to the January 1993 Jack-in-the-Box deaths in Washington State caused by contaminated meat, the ongoing debates about the safety of biotechnology, and the 1996 mad-cow disease scare in Great Britain as evidence that industrial agriculture is not without risk or high cost and is, in the long run, potentially disastrous.[8]

Though industrial agriculture still reigns virtually undisputed—every land grant university, agricultural scientist, and agribusiness chief executive officer assumes it, as do such powerful political figures as the secretary of agriculture and most members of Congress—its monopolistic power over American farming has recently been challenged by a loose collection of alternatives that share an "underlying philosophy" stressing "organic or near-organic practices" (Beus and Dunlap 594). These relatively new perspectives pay attention to the ecological and environmental dimensions of agriculture and the social and moral aspects of the national food system's impact on local rural communities. Alternative agricultures focus on sustainability; work to limit technological use and environmental damage; practice proven, low-input farming methods, whether ancient or modern; and promote crop and livestock diversity. In recent years the growing awareness of alternatives to prevailing farming methods has been coalescing into an alternative paradigm to the conventional, industrial view of farming.[9]

The conflict between these perspectives does not merely replay the old tension between agrarian and industrial values; the alternative paradigm, though supporting many "agrarian values," marks a break from previous agrarian versus industrial debates because of its focus on the ecological and environmental costs associated with industrial farming (Beus and Dunlap 595). Industrial agriculture views alternative farming as a danger and is working hard to see that its definition of agriculture will not dominate the twenty-first century. Will agriculture grow more large-scale, capital-intensive, and exploitative of nature—with an emphasis on fewer farmers and unlimited, vertically integrated expansion, as today's conventional paradigm argues it must? Or will agriculture do an about-face and stress small-scale operations, environmental responsibility, harmony with nature, more farmers, and checks on exploitative expansion, as the alternative paradigm hopes? The answers will determine not only what we will eat and how much we might pay for it; more importantly, they may determine whether we will eat at all.

In 1990 rural sociologists Curtis E. Beus and Riley E. Dunlap dubbed the emerging, loose collection of agricultural practices an "alternative agriculture" paradigm, which they argue is a real threat to the preeminent position of the industrial, "conventional agriculture" paradigm (590). To illustrate that competing paradigms exist, Beus and Dunlap compare the writings of twelve individuals well known to be concerned with the direction and structure of American agriculture. Six of these thinkers are apologists for conventional farming; the other six are proponents of alternative methods.

Beus and Dunlap conclude that "the two sets of writings reveal dramatically divergent perspectives on a wide range of agricultural issues" (590). They group their findings into six divisions that mark the gulf separating conventional and alternative thinking: centralization versus decentralization, dependence versus independence, competition versus community, domination of nature versus harmony with nature, specialization versus diversity, and exploitation versus restraint (see table). Beus and Dunlap acknowledge that the first three pairings describe a long-standing agrarian versus industrial debate, but they argue that the final pairings' focus on nature indicates the emergence of a fundamentally new farming paradigm. To support their argument, Beus and Dunlap compare the speeches of former secretary of agriculture Earl Butz with the writings of essayist, poet, and farmer Wendell Berry. Marking the "void" between the industrial and alternative farming paradigms, their views suggest a "schism in agriculture" (Beus and Dunlap 593).[10]

## "An Old Pattern of Entrances"

The histories of these men suggest their diverging perspectives. Butz never looked back after leaving his parents' Indiana farm and heading to college. After graduating from Purdue with a doctorate in agricultural economics, he worked his way through academe before being named assistant secretary of agriculture in 1954. He was dean of continuing education at Purdue in 1971 when he was nominated by Richard Nixon to be agriculture secretary (Blair). Before assuming his post and to avoid any conflict of interest, Butz resigned from the boards of directors of Ralston Purina, International Minerals and Chemical Company, Stokely Van-Camp, and Standard Life Insurance of Indiana ("Nixon Agriculture Nominee"). Suggesting the incestuous relations between government and agribusiness industry, he was replaced on the Ralston Purina board by an old student of his, the man he replaced as secretary of agriculture, Clifford Morris Hardin (Hightower 94–95; see Naughton). After his resignation following reports that

## Key Elements of the Competing Agricultural Paradigms

| CONVENTIONAL AGRICULTURE | ALTERNATIVE AGRICULTURE |
|---|---|
| **Centralization** | **Decentralization** |
| • National/international production, processing, and marketing | • More local/regional production, processing, and marketing |
| • Concentrated populations; fewer farmers | • Dispersed populations; more farmers |
| • Concentrated control of land, resources, and capital | • Dispersed control of land, resources, and capital |
| **Dependence** | **Independence** |
| • Large, capital-intensive production units and technology | • Smaller, low-capital production units and technology |
| • Heavy reliance on external sources of energy, inputs, and credit | • Reduced reliance on external sources of energy, inputs, and credit |
| • Consumerism and dependence on the market | • More personal and community self-sufficiency |
| • Primary emphasis on science, specialists, and experts | • Primary emphasis on personal knowledge, skills, and local wisdom |
| **Competition** | **Community** |
| • Lack of cooperation; self-interest | • Increased cooperation |
| • Farm traditions and rural culture outdated | • Preservation of farm traditions and rural culture |
| • Small rural communities unnecessary to agriculture | • Small rural communities essential to agriculture |
| • Farmwork a drudgery; labor an input to be minimized | • Farmwork rewarding; labor an essential input to be made meaningful |
| • Farming is a business only | • Farming is a way of life as well as a business |
| • Primary emphasis on speed, quantity, and profit | • Primary emphasis on permanence, quality, and beauty |
| **Domination of Nature** | **Harmony with Nature** |
| • Humans are separate from and superior to nature | • Humans are part of and subject to nature |
| • Nature consists primarily of resources to be used | • Nature is valued primarily for its own sake |
| • Life cycle incomplete; decay (recycling wastes) neglected | • Life cycle complete; growth and decay balanced |
| • Human-made systems imposed on nature | • Natural ecosystems imitated |
| • Production maintained by agricultural chemicals | • Production maintained by development of healthy soil |
| • Highly processed, nutrient-fortified food | • Minimally processed, naturally nutritious food |

*(continued on next page)*

Key Elements of the Competing Agricultural Paradigms (continued)

| CONVENTIONAL AGRICULTURE | ALTERNATIVE AGRICULTURE |
|---|---|
| **Specialization** | **Diversity** |
| • Narrow genetic base | • Broad genetic base |
| • Most plants grown in monocultures | • More plants grown in polycultures |
| • Single crops in succession | • Multiple crops in complementary rotations |
| • Separation of crops and livestock | • Integration of crops and livestock |
| • Standardized production systems | • Locally adapted production systems |
| • Highly specialized, reductionistic science and technology | • Interdisciplinary, systems-oriented science and technology |
| **Exploitation** | **Restraint** |
| • External costs often ignored | • All external costs considered |
| • Short-term benefits outweigh long-term consequences | • Short-term and long-term outcomes equally important |
| • Based on heavy use of nonrenewable resources | • Based on renewable resources; nonrenewable resources conserved |
| • Great confidence in science and technology | • Limited confidence in science and technology |
| • High consumption to maintain economic growth | • Consumption restrained to benefit future generations |
| • Financial success; busy lifestyles; materialism | • Self-discovery; simpler lifestyles; nonmaterialism |

*Source*: Curtis E. Beus and Riley E. Dunlap, "Conventional versus Alternative Agriculture: The Paradigmatic Roots of the Debate," *Rural Sociology* 55, no. 4 (1990): 590–616; reprinted here with permission of the Rural Sociological Society.

he made racist comments about blacks, Butz made much money from his notoriety through public speaking engagements.

Wendell Berry taught English at Stanford and New York University before choosing to return to Kentucky to become a professor, farmer, and writer. Arguing for a farmer-and-land-friendly agriculture, he took as his model the small, self-reliant homestead of the Jeffersonian ideal. An established poet and environmentalist, Berry worked for several magazines concerned with small farming and the environment, including *Orion* and *New Farm*. In 1977 he published *The Unsettling of America*, whose stated purpose is to answer "the assumptions and policies of former Secretary of Agriculture Earl L. Butz" (v).

In 1977 the two men debated their perspectives on farming at Manchester College, North Manchester, Indiana ("Earl Butz versus Wendell Berry" 50). Both left the confrontation admitting that neither had understood the other. In one of his rebuttals, Butz remarked, "I've got a feeling that Dr. Berry and I haven't met here tonight. Perhaps we won't" (56). Berry's reaction was similar: "Mr. Butz and I may never meet, because he's arguing from quantities and I'm arguing from values" (53). For example, summing up his understanding of agriculture during the debate, Butz lays out "Butz's Law of Economics—it's a very simple one: Adapt or Die" (53). And he acknowledges that "you make some trade-offs in this world. You lose some of the old family entity" (57). Whereas Butz's definition of farming is based on numbers, Berry's is rooted in people populating a place over time. He tells the story of a neighbor who envisioned buying a small farm but could not afford to after finding out that it would cost him $120,000 (54–55). For Berry, a farmer is not a "component of a production machine. He stands where lots of cultural lines cross" (55). To lose a farmer is to bury our cultural memory. A lot is at stake in these men's disagreement when we remember that farming constitutes 20 percent of gross national product and that every individual needs an uninterrupted, safe flow of food from ground to table (Easterbrook 3).

Berry recalls his debate with Butz ten years later in his novel *Remembering* (1988). Like *A Thousand Acres*, *Remembering* returns readers to the birth of the nation by nodding to the Declaration of Independence. In 1976 farmer and writer Andy Catlett attends a conference on "The Future of the American Food System" at a midwestern university. Invited to the conference as a token representative of those "who could hardly be said to be complacent about the Future of the American Food System," an angry Catlett answers—at high noon—the previous speakers, among them a "high agricultural official" who paraphrases Earl Butz: "I know there are some trade-offs involved in this. There is some breakdown in the old family unit . . . the basic law of economics is: Adapt or die" (22, 12).[11]

The novel offers an alternative vision to Butz's argument that farming is simply defined by the numbers. Andy Catlett knows that the high official will trade farm families for money: "The conference was about, and was meant to promote, the abstractions by which things and lives are transformed into money" (38). Catlett damns the conference as an "agriculture of the mind" that does not understand "how influence flows from enclosures like this [conference room] to the fields and farms and farmers themselves" (23). Catlett's words echo those of Vandana Shiva, who describes such thinking in her book *Monocultures of the Mind* (1993): "monocultures first inhabit the mind, and are then transferred to the ground" (7), erasing local agricultural knowledge in ways deliberately destructive of native cultures and land (9). *Remembering* reminds readers that such erasures are ongoing, destructive, and stoppable.

Butz and the high agricultural official do not believe that history matters; the past has nothing to teach and is best forgotten. The old farm days are "gone, and their passing is not to be regretted" (11; "Earl Butz versus Wendell Berry" 50). Defining farming as "big business," they know it only in the economic returns of the moment (12; "Earl Butz versus Wendell Berry" 51). In attending to the future they gloss over present losses, since, they argue, "problems with soil erosion and water shortages and chemical pollution" can simply be solved later (12). Both want to live in a "changing, growing, dynamic society" without defining their terms beyond the pursuit of "the amenities of life—color TV, automobiles, indoor toilets, vacations in Florida or Arizona" (11–12; "Earl Butz versus Wendell Berry" 50–51).

Answering Butz and other high ag officials, Berry has spent much of his life writing about the fictional farm community of Port William, Kentucky. In his depiction of Port William, Berry imagines a community whose literary roots stretch back to Sarah Orne Jewett's *The Country of the Pointed Firs* (1896). Jewett's "narrative of community" offers an alternative to Western literary preoccupation with the "individualized ego" (Zagarell 499). Narratives of community, such as *The Country of the Pointed Firs*, represent "a coherent response to the social, economic, cultural, and demographic changes caused by industrialism, urbanization, and the spread of capitalism." These works emphasize local daily life, community history, and community members' interdependence. Though predominantly practiced by women, narratives of community have been written by men who "felt some allegiance to, regional, rural, or working-class ways of life that were not emphatically individualistic" (Zagarell 499, 503, 512). *Remembering* is a contemporary example.

Berry reads *The Country of the Pointed Firs* as "the one American book . . . that is about a beloved community" ("Writer" 28). In his essay "Writer and Region," which appeared in the *Hudson Review* the year before he completed *Remembering*, he defines "the beloved community" as "common experience and common effort on a common ground to which one willingly belongs" (28). In his life's work, Berry offers an alternative male relationship with the land, one rooted not in Larry Cook's violence but in care. Just as Jewett imagined a narrative different from the linear male narrative of achievement, so Berry envisions a farming different from grid-defined conventional agriculture in his nonlinear, nonchronological *Remembering*. And though *Remembering* focuses on Andy Catlett, Catlett moves from self-pity to community reintegration, from leaving home to returning to it. Rather than rebelling against the village, Catlett embraces it. The "old pattern of entrances" that Berry represents in *Remembering* is in direct contrast to the entrances in *A Thousand Acres* (57).

*Remembering* argues that industrial farming is not an inevitability, but a choice. The Catletts' farming is an act of recovery. They restore to the culture's memory the possibility of a "pattern of a membership" that can be freely chosen (60). In coming home, Andy decides to reenter a community of mutual help, in contrast to the movement away from community represented in the conference attendees, who are "professional careerists of agriculture," and industrial farmer Bill Meikelberger, who has no neighbors and lives in a "deserted" house (23, 73–74). A national industrial economy argues that it cannot choose. For example, as the novel opens Andy dreams of a man in the basement of a bunkerlike building eating his own forearm: "I *have* to do this. I am *starving*. Three meals a day are *not* enough . . . This is my independence" (4). But Amishman Isaac Troyer offers another vision of independence. Having chosen not to buy out his neighbors, Troyer lives as part of a "healthy, comely, independent community" (84). In deciding not to mechanize his farm, he does no more than one "*ought*" to do (77).

Bill Meikelberger accepts a view of agriculture that is "as abstract as a graph or a statute or an airport" (81). Meikelberger, a "worried" man, admits, "You can't farm like this without having it on your *mind*" (74–75). Over the years he has carefully expanded his farm from its original eighty acres by "patiently buying out his neighbors" (73). A graduate of Ohio State's agricultural college, Meikelberger intends never to be debt-free: "Getting out of debt is just another old idea you have to junk" (75). Recalling Larry Cook, Meikelberger, a "Premier Farmer . . . one of the leaders of the shock troops of the scientific revolution in agriculture," has planted his two thousand acres entirely in corn (73). And much like Larry's, "Meikelberger's ambition had made common cause with a technical power that proposed no limit to itself, that was, in fact, destroying Meikelberger, as it had already destroyed nearly all that was natural or human around him" (76).

Andy has damaged himself in two ways: he sacrificed his right hand to a corn picker, and he spoke of his family to an academic conference that cared nothing for the personal (14, 25). He flees west in self-pity, but halfway through the novel, he decides to head back east. Knowing his home is "a place of history—a place, in part, the result of history," he rededicates himself to remaking a neglected farm that had been "diminished by its history" (95–96). In returning home, he reenters his and the nation's past to make right what was wronged: his family and the land. The Catletts' personal stories originally intersect when they are learning the nation's story; Andy and Flora first know their love for each other while studying for a history midterm that focuses on the "development and the influence of the mind of Thomas Jefferson" (111).

What has been done can be undone, with work, patience, care. Alluding to Dante and Milton, the novel begins in "darkness visible" and concludes with light, with Andy dreaming of the "infinite, sensed but mysterious pattern of its harmony," an alternative vision to that of the flesh-eating loner (3, 122; see Esbjornson 156). In the dream, he journeys on familiar ground with a guide who leads him to reimagine his neighborhood as a place defined by "the care of a longer love than any who have lived there have ever imagined" (123). By a "change of sight," Andy returns home to realize this vision of community; as he awakens, he raises the "restored right hand of his joy" to the living dead of his dream (124).

Flying home, Andy studies the grid that has defined American agricultural history: "The newer, larger roads are ruled according to the ideal of flight, deferring as little as possible to the shapes of the land" (104). From the air, in the "element of abstraction," the "details of the ground diminish, draw together, and disappear. The land becomes a map of itself" (102). Remembering details sheds abstraction and recovers one's lived experience, the love that is the world's only help (103). Back in Kentucky, on a road that follows "the shape of the country . . . curving along the bases of the hills, not like a pencil point along the edge of a ruler," Andy begins "to live again in the familiar sways and pressures of his approach to home" (117). Those he loves are now "presences, approachable and near" (117).

The tragedy of the agriculture represented in *A Thousand Acres* and *Moo* lies in the contemporary insistence that farming is a business and not a way of life. Rethinking farming as a way of life, as Berry does in *Remembering*, might foster a sense of community responsibility toward land, animals, plants—and humans, a responsibility sorely lacking in Larry Cook's life. Such responsibility—and the restraint it implies—is part of a "land ethic" expressed best by Aldo Leopold in his *Sand County Almanac*: "The land ethic simply enlarges the boundaries of the community to include soils, waters, plants, and animals, or collectively: the land" (239). Leopold notes that Americans' basic relation with the land has been solely economic, a fundamental relation that must change to include "love, respect, and admiration" (261). Including such qualities in the nation's farming might signal a change in how we perceive each other—leading us to think of each other as intrinsically worthy of interdependent respect and care. Completely erasing an economic relationship to the land may be impossible, but striving for a right use of land and imagining a more careful farming are ideals we must work toward. The future of the planet depends on this, this most georgic of ideas.

## "The Farm"

Most literary works that are critical of industrial agriculture couch their definitions of farming negatively: good farming does not abuse women and land, Smiley argues—thus defining her answer to industrial farming within the bounds of the conventional agricultural paradigm. To change how people think about farming, writers must capture readers' imaginations with representations of good farming that define alternative practices on their own terms, not as mere correctives to industrial farming. Only by doing so can the alternative paradigm take root in the culture's mind as a viable mode of thought and practice.

Wendell Berry's poetry embodies the careful artistry demanded by small farming. His latest collection, *A Timbered Choir* (1998), gathers several of his Sabbath poems written in the last twenty years. Berry's Sabbaths define farming as a community building act. He wrote each "in silence, in solitude, mainly out of doors" while on Sunday morning walks (Preface xvii). Though he cautions that they are not public statements, the poems do touch on "issues of public importance" such as the senselessness of war, the satisfaction of good work well done, the power of words well said (Preface xvii). For Berry, a farm is not only a subject but also a form (*Standing* 192–93).

"The Farm," a 428-line poem, remembers what industrial farming is sweeping from human imagination: a knowledge—and practice—that values human labor and intelligence, stresses cooperation with nature rather than opposition to it, and promotes responsibility in word and deed. With literary roots in Virgil's *Georgics*, "The Farm" imagines a diversity of people, animals, and crops rooted in a preindustrial paradigm that stands in stark contrast to the monocultural universe of industrial farming.[12]

First published in the *Hudson Review*, whose readers are primarily literary sophisticates, the poem addresses a single person, the reader, "you," a nonfarmer. Opening as an alternative, "The Farm" invites the reader-traveler to "Forget the wide road you / Have left behind, and all / That it has led to" and walk a "narrow road / Along the creek" (6–8, 1–2). Or "Best, walk up through the woods" to discover a clearing, a farm (9, 29). Brought to the farm "suddenly" as it lies "open to the light / Amid the woods," the urban traveler is offered the chance to make home and work one place (27–29).[13]

The vision that the poem imagines is not a pastoral one; this farm cannot simply be contemplated at one's leisure. It needs work and a care that deepens and broadens as time passes:

> Stay years if you would know
> The work and thought, the pleasure
> And grief, the feat, by which
> This vision lives. (44–47)

The poem defines good work as the best uses of nature needed to create a space for an interconnected human culture to thrive; as the poem reminds, "You did not make yourself, / Yet you must keep yourself / By use of other lives" (197–99). The reader confronts a choice not often offered in his or her present life: learn good work so you can do it yourself, so you can create your own space, or remain nostalgic about a place and practice others have told you is lost or archaic. Because people cannot imagine that an alternative exists:

> nowadays,
> A lot of people would
> Rather work hard to buy
> Their food already cooked
> Than get it free by work. (291–95)

As an answer to industrial agriculture, "The Farm" celebrates diversity by denying the power of monocultures, both of the mind and on the ground. Though conventional farming argues that crop specialization increases production, in the poem "by diversity / You can enlarge the yield" (171–72). To minimize the use of chemical fertilizers, "The Farm" advises alternating crops; the speaker explains how to sow a

> field in clover
> And grass, to cut for hay
> Two years, pasture a while,
> And then return to corn. (142–45)

The advice echoes Virgil when he says to "at another season sow with spelt / Fields you have stripped of beans" (Book 1, lines 73–74) and "Thus too by change of crops, fields can be rested / Without the thanklessness of untilled land" (Book 1, lines 83–84). Without proper work, without necessary diversity, disaster results. As "The Farm" points out, "The land must have its Sabbath / Or take it when we starve" (137–38).

For Berry and Virgil, only the small and the local offer sufficient space for community sustainability. Berry's farm is "Little enough to see / Or call across" (30–31); it is "a dear, small place" (416) that provides a "Beloved sufficiency"—though not necessarily a self-sufficiency (415). Such a place, "An independent state" (149), exists within a community of other small places in a neighborhood where one grows food, "some for yourself / And some to give away" (167–68). In contrast to industrial farming's elimination of neighbors and its exchange of food for money, Berry's farmer will "Eat, and give to the neighbors" (282).

Cyclic work and restraint creates the proper relationship between humans and nature. For example, when cutting trees each spring for firewood, the reader/farmer should

> Take the inferior trees
> And not all from one place,
> So that the woods will yield
> Without diminishment. (55–58)

Emphasizing the cyclic nature of good work through the poem's form, Berry returns readers to its opening when he closes with "You must saw, split, bring in, / And store your winter wood. / And thus the year comes round" (426–28). But this work has its restraints; like a tree, a farm "does not grow beyond / The power of its place" (230–31). The farm, like a tree,

rises by the strength
Of local soil and light,
Aspiring to no height
That it has not attained. (232–35)

Virgil also places the restraints of smallness and locality upon farm practices: "We must con its varying moods of wind and sky / With care—the place's native style and habit, / What crops the region will bear and what refuse" (Book 1, lines 51–53). And like Berry's poem, Virgil's is aware of the cycle of seasons—"For not in vain we watch the constellations, / Their risings and their settings, not in vain / The fourfold seasons of the balanced year" (Book 1, lines 256–58).

Rather than learning his craft from university experts, Berry's farmer learns proper work from the wilderness:

And while you work your fields
Do not forget the woods.
The woods stands by the field
To measure it, and teach
Its keeper. (214–18)

Nature is "the best farmer . . . Diverse and orderly" (219–23); a farm is "a human order / Opening among the trees, / Remembering the woods" (226–28). Trees teach that only unceasing work will maintain a human culture; without continually defining what separates farm from wood, an always present and ever ready wilderness will reclaim what was once cultivated.

Thus for Berry, a farm is not a thing; it is an event:

And so you make the farm
That must be daily made
And yearly made, or it
Will not exist. (342–45)

To keep the farm, one must work continuously and well:

There is no end to work—
Work done in pleasure, grief,
Or weariness, with ease
Of skill and timeliness,
Or awkwardly or wrong,
Too hurried or too slow

One job completed shows
Another to be done. (334–41)

Without ceaseless work,

This clarity would be
As if it never was.
But praise, in knowing this,
The genius of the place,
Whose ways forgive your own,
And will resume again
In time, if left alone. (348–54)

The work done on Berry's farm redefines "rich" in qualitative terms, a far cry from the quantitative denotation the word has in conventional farming; Berry's farmer builds soil—"The ground is mellow now, / Friable and porous: rich" (139–40). Berry's farm ethic grounds the land ethic that Aldo Leopold urged:

This is not work for hire.
By this expenditure
You make yourself a place;
You make yourself a way
For love to reach the ground.
In its ambition and
Its greed, its violence,
The world is turned against
This possibility,
And yet the world survives
By the survival of
This kindly working love. (202–13)

All this work makes the farm that makes the person who makes the farm: "And [the farm] is who you are, / And you are what it is" (369–70). This vision of farming positively reworks that represented in *A Thousand Acres*: "A farmer looks like himself . . . but he also looks like his farm. . . . What his farm looks like boils down to questions of character" (199).

Berry offers an alternative mode of thinking and practice that cannot be dismissed as anachronistic. He does not argue for a return to an old way of farming—the issue of feeding people is too important to be clouded in misty-eyed romanticism. The debate comes down to defining good farming and thus deciding who will control food production. What is the most responsible marriage of proven practices and new technologies and methods? Where should the power of food production lie: with farmers and consumers tied to each other in a local web of mutual dependencies? Or with centrally located, national and international agribusiness firms tied only to scattered stockholders? We have choices—and a lot of work to do.

**Work, then, is where we should begin.—Richard White, in *Uncommon Ground*, 1995**

# Postscript

## Fixing Fence

These words have a history. And placing them down defined me as much as it did my argument. It is clear now that the subject chose me long before I chose it; I can mark nearly to the day and hour when I realized that I had been writing all along.[1]

When I was growing up, it was pretty clear that I had a choice to make by the time I graduated from high school: go to college, move away and get a job, or stay home and farm. Most of my friends did not face this choice; they had no farms to worry about. They assumed that moving away for college or a job was what one did. I compromised. I went to college and returned home. Then I went to graduate school and came back a second time to farm. I admit, the place has me coming and going.

Our farm is nestled among the folds of the Appalachian Mountains in northeastern Pennsylvania, just north of the Poconos. We raise dairy cattle —never more than forty to forty-five head milking. I remember when we had sheep; my father remembers when there were chickens, pigs, horses, big gardens. Today, we specialize. In this there is nothing special; everyone who farms near us specializes in dairying. We scramble to make more and more milk and spend our money on food we should have grown ourselves. The competition heats up, some of us are burned off, and those left are left more alone.

We are all staring at the end.

Several years ago, I covered our township meetings for a local newspaper. At one meeting, the supervisors read a letter from state officials informing them that new restrictions were to be enacted limiting the amount of water used by township residents. I thought conserving water was a good thing, until I found out that these restrictions were part of a larger plan to conserve water in our section of the state for use in Philadelphia, a hundred miles to the south. The city was running out of potable water. Limiting ours would leave more in the Delaware Watershed to satisfy Philadelphia's water needs. It still seemed reasonable; I took a neighbor-helping-neighbor attitude. But it was further revealed that the state had plans to sink hundreds of wells all over the township, pump water, and allow it to run into streams, thus feeding the river. Someone in Harrisburg was paid to make these decisions. So far, the state has not implemented the plan, but ideas once laid down do not die.

There are 50,000 farms in Pennsylvania, and Pennsylvania is the fourth largest milk producer in the nation with sales totaling $1.46 billion. The sale of cattle and calves is the state's second most important income generator, totaling $522 million. Agriculture is Pennsylvania's number one industry, providing 600,000 jobs, generating 16 percent of the state's gross product, and pumping $538 million into the economy in export sales ("Why Should You Care about Agriculture?").

What do the numbers *mean*?

As a sophomore registering for the fall semester at a branch campus of Penn State, I found myself one credit shy of a full schedule. My choices: take a three-credit class and work hard or take a one-credit class and squeak through. The only one-credit course open that I could remotely connect with what I was doing in school was an agriculture course called "Careers in Ag."

What I was doing in school was trying to be a writer, and the only way—so I figured at the time—to be a writer was to be a journalist. I commuted to school an hour every day from our dairy farm, I was on track to be a journalist, so why not claim to be an ag journalism major? I was not sure that such a thing existed, but it sounded good enough to sign up. To keep my financial aid, I had to have a full schedule, and there was no way I was going to pay for three credits.

So I registered for this careers in agriculture course and showed up at the first class. The instructor was white-haired, almost perfectly round at the waist, and wore silver wire-rimmed glasses that masked green eyes. He was pretty uncomfortable-looking, and during class, to kill time, he talked of anything that came to mind. He arrived each day without notes, books, pens, paper, or ideas; he came with chatter, and all of us—four men and two women—just sat there and listened.

The first thing this instructor had us do that first day was to explain just what it was we were preparing to do *with* agriculture. The litany of careers everyone was anxiously pursuing centered on extension work, dairy science research work, and ag product sales. Our instructor nodded approvingly at each option as he strode back and forth before the blank blackboard, mumbling encouraging words about the job outlook for each and listing who he knew in that particular part of the "profession," recounting how great it was that agriculture provided so many wonderful opportunities. My turn came. Though I was still pretty unsure about my real plans, in the circumstances I felt obliged to recount my reasons for taking the course. My career choice was ag journalism—and, after all, I kept reminding myself, journalism was my major. Quizzical looks crossed everyone's face, and I stammered to explain that I meant to become an editor of a farm paper. Our instructor frowned; he had been expecting all technical-career-minded people. After quickly counting the number of students in the class, he said I could stay, and he did mention in passing that the industry always needs people to keep consumers *properly* informed.

Henry Gordon, our instructor, was the local county extension agent—an expert on "the dairy industry." His columns on dairy prices and trends in the industry appeared regularly in regional farm weeklies. So his name was familiar when I signed up for the course: I had read his arcane prose and slogged through his densely worded articles to find that things ran swimmingly in the industry as a whole, "though some parts might be momentarily struggling."

Nothing technical went on in Gordon's classes; he lectured about men—no women—he knew who had made it good in agriculture. None of those who had made it good were farmers, curiously enough. Every anecdote centered on what everyone, other than myself, was attempting to break into: research, sales, extension work. Gordon felt under siege in the classroom, almost as much as we felt besieged by him—that much was obvious. The best alternative he could think of was to leave the classroom, and that is what we did on a regular basis: we took field trips in the most literal sense. We went to cornfields, hayfields, tomato fields.

We took our first trip on a sunny, September day—harvest time. Gordon assured us that we would see the whole tomato-harvesting process: from field to crate to packaging to marketing to sales. Somehow, I ended up sitting in the front seat between Gordon, who was driving, and a beefy freshman majoring in dairy research who kept drumming his fingers on the dashboard. Irritated by the racket, I was more talkative than usual, and I asked Gordon how he felt about the price of milk farmers were getting and what was going to happen to it in the near future. After all, he was the expert.

Just that morning, after we had finished milking, my parents, and the rest of us in the family, had had a long, loud, complaining argument at the breakfast table about less and less money coming in for ever-rising bills. Selling out suddenly loomed as a possibility. One of the bills happened to be a tuition bill, which my father spilled milk on, only to fan another uproar. I really felt like apologizing that I had gone to school in the first place—and to be a *writer*. Should I drop out? None of this was new at home, but it was all recent enough to cross my mind as Gordon guided us beyond the city.

Gordon quickly warmed to a subject about which he had something to say and replied that milk prices were going to drop further and a good thing it was so we could get rid of marginal, inefficient farms. Keeping the most profitable farms going, getting them bigger, making them even more profitable, was the key to a "healthy industry." I took this in, staring out the window at a row of fast-food places jerking by—Gordon had the nervous habit of pressing and releasing the gas pedal every few seconds. I finally turned to him—recalling again that morning's breakfast conversation—and asked this industry expert about the people who might be forced out of business. What will they do? My mind, of course, was on my family's situation: I did not want the price of milk to drop any further, endangering *our* farm. Pulling on the steering wheel, Gordon wriggled himself into a straighter sitting position and told me that I was taking it all too personally, much too personally. The industry cannot afford inefficient farms. You cannot look just at what happens to individual farms. Look at the big picture. Putting small, not-so-profitable farms out of business is good for everyone. I was about to reply that it was not so good for those who are sold out, but we were passing a turkey farm—a turkey factory farm—and he reminded us all that if we wanted a good, fat turkey for Thanksgiving we had better get our order in now. Anybody want to stop?

No.

I was the only one in the car who lived on a farm, had been raised on a farm. And that included the expert in the driver's seat. The others wanted to be part of "the industry's big picture": research, sales, extension work. A few, and it was incredible to me, had never set foot on a farm, not even the beefy freshman still drumming on the dashboard. They would be part of Gordon's big picture, working in something that they had no firsthand knowledge of, making decisions, forming policies, and creating mind-sets that trickled down to affect, finally, individuals on individual farms. But, as they continually claimed, they liked animals.

The man who owned the tomato farm was president of our state's farm association. He was tall, thin, and wore a blue cap emblazoned in red with his association's name. His expression and countenance implied that he would at any moment throttle the nearest living thing. We toured his tomato operation with reverence. Inside one barn we sat in what was for me a curiosity: a farmer's office, a big one, wall-to-wall carpeting, walls adorned with pictures of Ronald Reagan, George Bush, and a few senators and congressmen, all surrounding his Dickinson law degree. "Sales" and "Marketing" were stenciled on the door's frosted glass. Ray Thorn was one of the leading lights in an association of "industries of agriculture," in the forefront of helping to preserve "family farms," as his association's newsletter proclaimed to its "thousands of agribusiness readers." But I soon understood from his conversation that the farms his organization was preserving were remarkably like his own—and nothing like ours. We had no office, and the pictures we had were mostly of each other and our dog.

Standing outside in the September sunshine, beside the blacktop road, across from the well-groomed grounds of Thorn's two-story Victorian, I heard a cow bawl. Gordon and Thorn were studying the tomato packing going on in yet another barn a few hundred feet away, pointing to crates of green tomatoes and the blank-faced workers packing them behind a rickety conveyor. I had to wait until we were moving toward this barn, and their conversation had lapsed, before I could ask Thorn what he thought of the current dairy situation. Horrible, he replied good-naturedly. He went on to admit, with a touch of regret, that he kept the hundred or so dairy cows he had for sentimental reasons: to keep his elderly father happy and off his back. More, he related, the dairy part of his operation was running at a loss—as a matter of fact, it had been for two years now.

Later, it was rumored that this same state farm association president had sold his herd at an enormous profit several days before it was announced that the 1985 Farm Bill would include a herd buyout provision that would allow dairy farmers to sell their cows to the government as long as they agreed to stay out of dairying for five years. Of course, the price of cows dropped when the buyout was announced, but Thorn had already made his deal—convenient.

We toured Thorn's tomato-packing operation, and the agent and the association president had a long discussion lamenting the fact that no one had figured out yet what to do with culled tomatoes, useless by-products of the harvest that were simply tossed away. Maybe some of our young geniuses here will figure it out, Gordon said, winking at Thorn, but Thorn was leaving us, mentioning as he strode away that we might want to drive around looking at the fieldwork.

We found several picked-over tomato fields before coming to a remote field that was being picked over as we pulled up. We stood at the field's edge and watched migrant workers work. It was stoop labor. Each picker filled a bucket full of green tomatoes, lugged it across the field, and handed it up to someone standing in the back of a truck. The tomatoes were dumped onto a growing pile, and a quarter dropped into the bucket as it was passed back to the picker, who went further afield for more tomatoes.

Gordon touted this as a great way to make money. Want to make a buck? Just grab yourself a bucket and pick a few tomatoes, you will see that in a real short time you will make some good money. It is not hard work. You even get to work outside all the time. And it is a nice day. Think of it: on this job you get to travel up and down the East Coast, following ripening crops, you know. So you are always in the sun. You winter in Florida—how about that? We all watched, detached, as the pickers bent for more tomatoes, pulling back vines and tossing green tomato after green tomato into green buckets. The only conversation was at the truck: a man spoke Spanish to each picker as he or she handed up a bucket. The workers were young, old, men, women—families, someone murmured. There were no takers when Gordon repeated his offer to give us work in the field. And he laughed at our refusal. A lazy-kids-of-today look crossed his face as he turned toward his Dodge.

Gordon's Dart weaved its way down a hill, passed several shacks and long-abandoned equipment—rusting balers, broken rakes, stripped tractors. His chatter was nonstop now until we reached the main road. He praised the efficiency of Thorn's operation and all that Thorn was doing to help agriculture. He reminded us how many opportunities agriculture was holding out for us. All we had to do was reach out and take them, just like so and so did. . . . He went on and on. It did not register until after we had gone another few miles along the highway in silence that those shacks we had passed were where the tomato pickers lived when they were not working for those quarters. I was about to mention this to Gordon as we passed Thorn's home, but maybe such an observation would be too personal.

Penn State University, of course, is fed by legions of industries hovering on the periphery of agriculture: petrochemical fertilizers, large equipment manufacturing, genetic engineering. Many of the school's ag research innovations and discoveries require changes in equipment or practice that force farmers to rely more and more on off-farm businesses just to break even. An extension agent has a vested, personal interest in keeping this vortex roiling. On-farm self-reliance is washed away, along with the better part of each person. Farmers now see this as a ring of natural interconnectedness, rather than a drain, and do not question the vicious cycle they spin within, at least not until its centripetal force sucks them out of the "industry" as one of the unprofitables.

The connection between the university researcher, working out and thinking out agriculture's larger problems, and the farmer working out and thinking out agriculture's local problems has been smashed. The symmetry that was—if it ever was—is broken. Any individual-to-individual connection has vanished. Now, it is group-to-individual. Land grant universities define extension in its most linear sense—one way—to peddle blanket solutions to every garden bed problem. The latest example: farmers plant not by sight but by satellites, using computers mounted on tractors. With their heads in the clouds, there is no need for farmers to see the ground under their feet. The farmers' web of dependence is no longer global; it is extraterrestrial.

In November 1989 the Berlin Wall was torn down, piece by piece, by people it had separated physically and politically. Germans who had been East Germans only a month or so before soon flooded into West Berlin to taste a freer political system. Scenes of East Germans strolling the streets of West Berlin, wringing the hands of West Berliners, exulting in the change in political weather dominated our kitchen TV. A weight lifted from Europe right before our eyes. What was going to happen?

At the same time, announcements of failed savings and loan companies broke the silence of the morning's first cup of coffee. Government scandals, drug problems, judicial breakdowns flooded across the kitchen table in the near dark. Every once in a while during these reports I would get the feeling that something was unraveling. Government on the home front did not look good.

One day after a dose of these conflicting images, I received a telephone call from the editor of a local newspaper. Could I cover a nearby township meeting that night? I had been sending him stories on our own township and community center happenings. The meeting was at eight o'clock at the township secretary's home. Sure, I said.

I arrived first. The secretary and her husband were just finishing coffee and cake, and I made myself as inconspicuous as possible on their couch. An Irish setter jumped up beside me. The husband said little, grumbled something about the "goddamn township," and disappeared into another room.

"I'm Jean," she announced as his footsteps ascended a stair. "You must be Billy Conlogue's boy."

"Yeah," I replied, smiling at the greeting I usually encountered at area farms. The dog pawed my notepad.

"I know your dad. You look just like him."

I answered that I never knew whether that was good or bad, and she laughed and rummaged in a kitchen cabinet.

"The others ought to be along soon," she said, pulling out a leather case that evidently held the township's business.

She shooed the dog outside, and I glanced at the clock between the two windows across from me. It was 8:15.

Ten minutes later a face looked in the back door and Walt Ridge stepped in, slipped off his jacket, and sat at the dining room table. He had been in the buyout five years ago and was about to start up again with cows; he had not seen a winter so calm as this one was so far and how about all that happening over in Germany?

"It's something," Jean answered, handing him a piece of paper. "Did you buy this oil filter at Burton's garage? He sent me two bills on it."

Ridge slipped on his glasses, tilted the pink slip toward the light overhead, and looked it over carefully.

"I bought one and Rob musta bought one. Got mixed up on who was to get what."

"Well, I'll pay for both, but you two get your signals right next time. We don't need a double order of salt or cinders."

The clock chimed 8:45.

"Rob and Harvey'll be late," Walt said, as if the chiming had reminded him. "Harvey's milking machine went on the fritz again."

"Rob took his old floor models up there?" Jean replied. "No telling how long they'll be."

The couch was too comfortable, and at that announcement I moved to the high stool beside the china closet. It looked like this might take a while. But, just before nine, Rob and Harvey could be heard coming along the walk, scuffing their work shoes on the concrete and complaining loud about the damned Delaval dealer over in Jeffersonville never wanting to come out at night and then never bringing the right tools.

The door swung shut, and Rob and Harvey and Walt discussed Harvey's milking and made predictions about how it would go in the morning, what with Harvey's son gone hunting up in New York State and his daughter at college.

Rob cleared his throat and the conversation ebbed. Jean introduced me to be polite.

"Conlogue from over in Pleasant Mount?" Walt asked, peering at me over his glasses.

"Pretty stony ground up there," Rob said.

I nodded. I mentioned that I covered the meetings for our township, and Harvey turned to me.

"That's where we got that old truck a' ours from," he said. "A '73. Runs but burns oil like there's no tomorrow."

"They got a new Jimmy this year, didn' they?" Walt asked Rob.

"Yeah, wouldn't get another Jimmy if they was givin' it away."

They lapsed into silence at the thought of their last GMC.

"I seen that damn truck down in front of Carliss's in Honesdale, stripped to a shell." Walt shook his head, whether from regret or quiet satisfaction, I could not tell.

Jean shuffled the bills and Harvey nodded. Walt and Rob were newly attentive.

"This meeting's now to order," Harvey announced rather formally. "Let's have the minutes of the last read."

He turned to Jean and she read the account of the last meeting, asking Harvey here and there if he had done this or that since last month; invariably he said he had. Once in a while I caught Walt and Rob exchanging glances, and it was clear to me that Harvey was being reminded of things he needed to do. Jean sighed and continued on after each interruption. With the bills read, she pulled out the township's accumulated correspondence.

"It's mostly bills, of course," she said. "Here's an invitation to a seminar on road maintenance down in Scranton, but nobody wants to go to that do they, we know enough about keeping up the roads, don't we?" She looked at Walt.

"We better by now," he said. He leaned back in his chair. "Ain't heard any complaints lately anyhow, unless you count Sally Tressle."

"Now what's her problem this time?" Rob asked, leaning over the table. "We cindered her road first last time."

"I know it," Walt answered. "But she claims the cinders blowed off."

"Ah," Rob replied, disgusted. "If it ain't dust she's complainin' about it's cinders."

Jean reined in the conversation by pulling out another letter.

"Got a thank you here from Dairylea for pulling out that truck after the last snowstorm," she announced.

"Fool driver tryin' to make West Hill so empty in so much snow," Harvey commented. "Same new guy almost backed into my milk house this mornin'. Who the hell is he anyway?"

"Isn't he Drinker or Dinker or Dinkens?" Jean offered. It seemed important to know.

"No, no," Walt answered, emphatically shaking his head. "He's Kinneren. Bob Breaker's son-in-law."

"That oughta explain a few things." Rob smiled.

Jean opened up and tossed out the half-dozen remaining letters, most of them advertisements of trucks that they could not afford or announcements of meetings and seminars that they had no interest in attending. They knew their business. Then, she turned to the bills.

"Now, I already talked to Walt about one bill," she said and continued on, not stopping for the puzzled looks crossing Rob and Harvey's faces. But the looks vanished in shrugs of trust.

Catching that simple action, I was suddenly aware of something breathtakingly obvious—these four individuals were a government. Duly elected, they were keeping this corner of the world for their neighbors. They knew their place and the people in it. And I imagined that all over the country, men and women—ordinary folks—were going about this same task on the first Monday of the month or whenever, coming together to make sure that things were running well and that the bills were paid. They were practicing care; those shrugs of trust suggested a faith that too many federal and state bureaucrats had forgotten.

I have since thought that if our Congress would do what we did after that meeting, the world would be a better place. When the meeting was over, we cleared the table and sat in Jean's kitchen eating ice cream and having coffee. And we talked, just talked.

One Sunday in August, I was sitting at our kitchen table studying *Walden* when my brother Danny interrupted me. Tall and wire-thin, he loomed over me with a cigarette dangling from one corner of his mouth.

"I want you to get up the hill and put a fence along the road, so I can get the cows in the meadow."

"Uh-huh." I kept reading.

"Do it now," he ordered, stripping me of the book. "I want those cows in the meadow after supper."

I shrugged and leaned back in my chair. Home from graduate school for the summer, I did not argue. "I'm supposed to put up half a mile of fence—alone?"

"Yeah," he replied, glancing at the book before tossing it aside. "Use that ribbon fencing we got."

"Never used it."

"It's easy. Stuff's in the garage; there's a hammer on the back porch."

"All's I need's a nail hammer?"

"That's it."

"What're you gonna do?"

"Watch the Phillies."

I do not like ribbon fencing, an inch-wide strip of polyester interwoven with tiny threads of steel. Hard plastic posts, about four feet in length, hold it up. Each post can be tapped in the ground with a nail hammer; no twelve-pound sledge is needed to build this fence. It is supposed to be great for creating temporary pastures, such as marking off a meadow as I had to do that afternoon. It was a snap. In less than two hours, I had that fence up, a pasture ready to go.

When I finished, I admit that I appreciated the ease with which I had erected it. In no time flat, I had set up a few hundred or so posts, rigged up a sun-powered charger, and unrolled enough ribbon to enclose acres of meadow for our herd's late summer grazing. I was not sore from pounding posts, and I had not cut myself on barbed wire. I thought I might even be able to see the late innings of the game. Ironically, it was the ease of construction that bothered me most. The fence was virtually ready-made, simple, clean, without hazard. I had so little to do with its making that I felt uneasy when it was finished. Along the road, as far as I could see, the ribbon billowed gently in the August breeze, a white strip waving in the meadow's dull green. It looked remarkably flimsy.

I stood there and thought long and hard about why I would rather have spent all day building the usual barbed-wire fence. Though it was hot—the sun was merciless—I remained rooted at the end post, trying to make sense of my reaction. In making the fence, I had no need to dig post holes, or to make sure the posts were spaced properly; I did not even have to stand the posts straight. Nor did I have to pull the ribbon tight. Constructing such a fence is a careless act: it requires no concentrated attention, no effort of imagination, and no one's help. My only choice in building it was whether or not to turn on the charger. I flipped the switch.

A week after the cows went into the meadow, the days turned cloudy, so the charger did little charging. Without electricity, the fence was just a ribbon; the cows soon sensed this and were out, pulling ribbon and posts down the road with them. Suddenly, there was no pasture; the meadow was back, and there was the world for the cows to run in. They leveled a cornfield, tore up a neighbor's garden, and punched up another's yard. Getting them back was no easy matter; the only thing easy was remaking the pasture. But once a cow has breached a fence, she will keep trying it, no matter how many times you set it up. Seeing no distinction between one place and another, she is ready and willing to wander away.

The cows' August rampage happened to interrupt me while I was finishing *Walden*. When I left the house, I dropped Thoreau as he was leaving his pondside home.

By the time I picked up Whitman, the cows were out again.

Fixing fence is a spring ritual. In early April, after frost leaves the ground, we fix fence, my brother, my father, and I. We straighten posts, pull wire taut, splice old wire to new. Fences are marked with histories of long winters. Repairing that history takes thought, imagination, and others' help. It takes effort to maintain a good fence, especially a line fence.

It is midmorning. The cows are fed and bedded down in the barn. It is cool, so we will not have to worry about mayflies. I wait on the tractor seat; Danny and my father check the number of posts and staples in the trailer hitched to the tractor. They make sure we have enough wire.

The posts are locust, cut from a stand planted by someone with foresight who farmed here long ago. We cut the trees last fall and shaped them into posts yesterday. The new wood is lighter than the thick bark; edges and slivers stick up here and there. We handle each post with care.

Grass is greening—last winter's snowfall and the spring rains promise a lot of hay. The sun gets warmer, and my father takes off his light jacket before we start fixing. Like we do every year, we fix to the corner where the line joins Lindens' place. Danny handles the sledge hammer and pounds posts, my father splices and stretches the wire, I hammer staples and drive the tractor. We work, we talk, we complain.

Mrs. Linden fixes no line fences. When her husband died several years ago, she sold their cows. She was a nurse until she retired. Since then, her pastures have grown to brush; she rents to no one.

In the woods marking the bounds between her property and ours are set-in posts, sections of telephone poles placed below the frost line, standing six feet above ground. They are there to stay; along with the wire, they mark the exact bounds of our properties. Years ago, while I was still a child, Mrs. Linden's husband had taken it upon himself to redefine things; he rebuilt the fence several feet beyond where it had always been. To settle the dispute, surveyors redrew the line. Now, every time I work in those woods, I wonder how tempestuous the feud was that led to a day's work putting up those set-in posts. My father does not talk about it much. If asked, he will only mention that there had been a disagreement, and now it was settled.

Good fences make good neighbors.

One year Danny decided not to rebuild the line. When a fence is left to slacken and fall, boundaries become fuzzy. No one knows where things begin and end; you cannot tell if you have wandered away from one place and into another. The set-in posts became no more than sticks, purposeless. I cannot find the line anymore. Declaring difference does not help; differences must be shaped and shared. And that takes work.

I began rebuilding our farm's stone walls after I returned from Penn State, halfway to an undergraduate degree in English. Quitting school was a difficult choice. My father was ill with coronary heart disease, and our farm needed tending. My older brothers had their own jobs and families, and Danny was only thirteen. It was my place to fill while my father had a triple bypass. I came home.

He had his operation in November, and building walls was a kind of therapy that fall. Building was rebuilding: away a student, suddenly I was back a farmer. The first walls I reconstructed were retaining walls that rounded off the end of our cow barn and protected one side of our calf barn. Most of the rocks were scattered in the barnyard, and cows frequently bruised their feet on them, so clearing the barnyard and remaking the walls solved several problems.

"What are you doing?" my father asks. He has been home from the hospital only a month. His farmer tan has melted away; his clothes hang loose. He moves carefully.

"Building a wall; what's it look like I'm doing?"

We stand in the barnyard, he with a cigarette in hand, I with a rock. He looks out across the barnyard, at the litter of stones, the tufts of tough grass, the briars and nettles growing along the calf barn. He puffs the cigarette; he refuses to quit.

"Why're you doin' this?" he wonders.

He is not interrogating me, I finally understand.

"I thought it'd help smooth the barnyard, keep the cows from bruisin' their feet, keep the mud down."

"Uh-huh," he replies. He blows a cloud of smoke and looks out across the pasture at the walls separating the woods from the meadow. "I remember my grandfather used to talk about keepin' up the walls around. Said it was good work for a man to do."

"Yeah?" I lay stone on stone.

"I'm glad you're doin' it. Puts things to right."

"Uh, thanks." I thought so, too.

He leaves, and I am left to work, piling more stones. No one minds that I do this between caring for the cows and cleaning the barn. I keep the farm going, and the rebuilding goes on.

Soon there are new walls here and there, like sentences in the margins. People notice, ask questions.

"Why don't you use mortar on those rocks?" someone asks. "Wall'll last longer."

I shrug. I had thought of it, but using mortar struck me as asking for too much permanence.

"Looks real nice," an aunt says.

I smile. With the stones in place, it is the barns most people see differently; they notice a change, but they cannot quite say why.

I did rebuild a few of the stone walls around the house. Where before the yard's border had been indistinct, it is now clear; you can see where the pasture ends and the yard begins. Neither yard nor pasture, the wall gives expression to both. A few years later, with my father back farming, with Danny a little older, and the place in shape, I went back to school, back to piling book on book.

Soon I was piling word on word.

# NOTES

## Introduction

1. According to Marty Strange, "an industrial agribusiness system produces farms that are: *Industrially organized*," "*Financed for growth*," "*Large scale, concentrated*," "*Specialized*," "*Management centered*," "*Capital-intensive*," "*At an advantage in controlled markets*," "*Standardized in their production processes*," "*Resource consumptive*," and "*Farmed as a business*" (36–39).

2. For more on the pastoral, see Alpers, *What Is Pastoral?*; Ettin, *Literature and the Pastoral*; and L. Marx, "Pastoral Ideals," 252–57.

3. Other classical Greek and Roman figures who wrote about agriculture include Hesiod, *Works and Days*; Cato, *On Agriculture*; and Varro, *On Agriculture*. Cato was widely read by eighteenth- and nineteenth-century farm reformers.

4. There is only scattered attention to the georgic in American literary scholarship. For examples, see Altherr, "'The Country We Have Married'"; Gentilcore, "American Georgic"; and Tillman, "The Transcendental Georgic in *Walden*."

5. Literature on American agrarianism and the yeoman farmer is extensive. See, e.g., Johnstone, "Old Ideals versus New Ideas"; Hofstadter, *Age of Reform*; H. N. Smith, *Virgin Land*; and L. Marx, *Machine*.

6. Wayne E. Fuller points out that many leaders of the national Progressive movement "owed much to rural America for those dominant traits which were so characteristic of them" ("Rural Roots" 3).

7. See chapter 3.

## Chapter One

1. The word "bonanza" also describes large deposits of gold and silver.

2. I thank Lewis Lawson for this insight.

3. Bigelow notes: "Throughout my tour it was noticed that there was a great abundance of unemployed labor" (41).

4. For another argument about the abstraction of nature in *The Octopus*, see Sarver, *Uneven Land*, 75–103. For the novel as "corporate fiction," see Michaels, *Gold Standard*, 183–213; for its relation to an "urban sublime," see Den Tandt, *Urban Sublime*, 70–82.

5. All page references to *The Octopus* are from Frank Norris, *The Octopus*, afterword by Oscar Cargill (New York: New American Library, 1981). Norris appears to have had exact knowledge of the Mussel Slough area. The two dry seasons mentioned in the novel may refer to Mussel Slough's "protracted drought of 1877–1879" (Preston 91). That Magnus Derrick is a North Carolinian reflects the area's "heavily Southern" demographics (Preston 75).

6. Norris's descriptions of plowing recall Coffin's illustrations: "The plows, thirty-five in number, each drawn by its team of ten, stretched in an interminable line, nearly a quarter of a mile in length. . . . They were arranged, as it were, in *en echelon*. . . . At a distance the plows resembled a great column of field artillery"

(94). Annixter says of Broderson: "Fool! . . . Imagine farming a ranch the size of his without a foreman" (89).

7. Historically, miners like Magnus Derrick, who were "familiar with the techniques of water diversion developed in conjunction with hydraulic mining," were the first to farm "in earnest" in the Tulare Basin region, of which Mussel Slough is a part (Preston 75).

8. See "An Appeal to the People," *Visalia Weekly Delta*, 7 May 1880; "A Collision," *San Francisco Chronicle*, 12 May 1880; and "Then and Now" and "Pioneers in Mussel Slough," *Visalia Weekly Delta*, 4 June 1880. Portions of these articles were republished as a pamphlet, *The Struggle of the Mussel Slough Settlers for Their Homes!* See pp. 14, 16, 21, 30.

9. The idea of property rights is itself a European imposition on the landscape, in contrast, for example, to Native Americans' more communal conceptions of the land.

10. For Jefferson's survey and its connection to earlier American literature, see P. Fisher, "Democratic Social Space."

11. David Wyatt notes that "*The Octopus* begins with a man mapping out a landscape. . . . Presley's bicycle ride accomplishes two clear narrative functions; it acquaints us with the territory and introduces most of the central characters" (108).

12. Others argue that Vanamee ends up simply deluded, if not insane. Barbara Hochman claims that Vanamee's "conviction at the end of the novel that he has come to prove 'Time was naught; change was naught . . . Death was overcome' . . . involves a mad belief that he has literally transcended 'immovable facts' " (33).

## Chapter Two

1. Cather's choice of paper and type for the original publication of *O Pioneers!* suggests her efforts to manufacture this nostalgia: "The heavy texture and cream color of paper used for *O Pioneers!* and *My Antonia* . . . created a sense of warmth and invited a childlike play of imagination, as did these books' large dark type and wide margins" (Rosowski and Mignon ix). In preparing their scholarly edition of *O Pioneers!*, Susan Rosowski and Charles Mignon "deferred to Cather's declared preference for a warm, cream antique stock" (x).

2. Recently, scholars have been studying Alexandra's business practices to support a variety of arguments. See Reynolds, *Willa Cather in Context*, 56; Motley, "The Unfinished Self," 150; Gustafson, "Getting Back to Cather's Text," 152; Urgo, *Willa Cather and the Myth of American Migration*, 45; and Horwitz, *By the Law of Nature*, 220–21.

3. All page references to *O Pioneers!* are from Willa Cather, *O Pioneers!*, edited by Susan J. Rosowski and Charles W. Mignon (Lincoln: University of Nebraska Press, 1992).

4. Carl claims that the Bergsons have become so rich that "Morgan himself could n't touch [them]" (105). Populists damned land speculation in their 1892 Omaha Platform: "The land, including all the natural sources of wealth, is the heritage of the people, and should not be monopolized for speculative purposes" (Hicks 443).

5. Rural sociologist Llewellyn MacGarr repeats the often noted claim: "In many cases of insanity among wives of farmers, the particular strain which brings about

the breakdown has been found in monotony, isolation, and overwork. The automobile and rural telephone should do much to relieve this particular strain" (96). MacGarr, *The Rural Community*, 182. See "Why Young Women Are Leaving Our Farms," *Literary Digest*, 2 October 1920. In *Barren Ground*, Geneva Greylock suffers a mental collapse that leads her family to contemplate committing her to an asylum (376; see 437–38).

6. Eugene Davenport defines three stages in the development of American agriculture: (1) "primitive agriculture . . . the self-sufficing system"; (2) the "money-making stage . . . at the expense of virgin fertility . . . that generation was exploiting nature at a rate never before attempted"; and (3) the "scientific stage of farming . . . the object is not so much the magnitude of production as it is the quality of the product and the economy of its production" (45–46). Herbert N. Casson defines the old farmer as a "Robinson Crusoe of the soil" and the new farmer as a "suburbanite" whose "business has become so complex and many-sided" (598). But, according to Casson, "All American farmers, of course, are not of the new variety. The country, like the city, has its slums" (598).

7. The Country Life Commission and the Commission on the Conservation of Natural Resources "were created simultaneously in 1908" (Ellsworth 161).

8. The University of Nebraska formed the Department of Agronomy and Farm Management in 1909 (Manley 202).

9. The extension system was consolidated as the Agricultural and Home Economics Extension System in the 1914 Smith-Lever Act (Bowers 89; see Jellison, *Entitled to Power* 16–17). For farmers' resistance to rural school curricular reform, see Danbom, *Resisted Revolution*, 77–78, and Fuller, "Making Better Farmers," 166–67.

10. For a photograph of the type of header Amédée is likely using, see Collins, *The New Agriculture*, 301.

11. For a full discussion of "the machine in opposition to the tranquillity and order located in the landscape," see L. Marx, *Machine*, 18.

12. The Nebraska Hospital for the Incurable Insane was built in Hastings in 1887 (Stouck 339). For Cather's unfavorable view of Populists, see Cherny, "Willa Cather and the Populists."

13. The story of his grandfather should remind Lou of what happens to those who indulge in "every sort of extravagance" (29). The old man's "unprincipled wife warped [his] probity" and encouraged him into foolish speculation that caused him to lose his fortune (29).

14. Most Populists supported the free coinage of silver (McMath 183). Nebraska Populist William Jennings Bryan opposed the gold standard and backed free silver (McMath 201). Wall Street investors and bankers endorsed the gold standard (Goodwyn 12).

15. Antipopulists often portrayed Populists as "the political expression of the misfits and failures" (Clanton 565). William Allen White editorialized in 1906 that the "'basis of [Populists'] contention was the envy of wealth—the hatred of the rich'" (Clanton 579).

16. All page references to *Barren Ground* are from Ellen Glasgow, *Barren Ground* (San Diego: Harcourt Brace Jovanovich, 1985).

17. Another reviewer claims that Dorinda was "created for no other purpose than to become a model dairy farmer and to act as demonstrator of the gospel of sci-

entific management in the redemption of the land that had been abandoned to elemental evil in the form of what the Virginians call 'broomsedge'" (Brock 247).

18. Blair Rouse notes that Glasgow "thought of the story as stretching from 1894 to 1924" (87). Godbold reminds us that "Dorinda's life coincides with that of the novelist" (138). In her autobiography Glasgow points out that the novel took "form and substance in [her] imagination" during and just after World War I (*The Woman Within*, 241). It is also important to remember Glasgow's urban orientation: she spent her life living and writing in Richmond, Virginia.

19. This marriage is particularly important when one considers that, in much of the South, "the modern world of the college of agriculture and the county agent was alien" until well into the 1930s (Kirby 49). Jason Greylock tries to promote progressive farming, too, but he does so smugly and without the practical farm credentials that Nathan displays (32). The weak, lazy, and vacillating Jason advocates theory without demonstrating its value.

20. As late as 1935, only 10 percent of American farm families received "central-station electrical service" (Beall 790). The 1920 census showed "that southern farmers had hardly begun to enter the age of telephones, automobiles, motor trucks, tractors, and electricity" (Fite, *Cotton Fields* 101). In 1920 only 1 percent of southern farmers had a tractor and 12 percent had an automobile (Fite, *Cotton Fields* 102). According to Fite, "There were still many southern farmers in 1919 who did not have a single milk cow and very little if any other livestock" (*Cotton Fields* 97).

21. For example, Rouse claims that "the interlude in New York when Dorinda conveniently suffers a miscarriage is out of harmony with the rest of the novel" (92). McDowell declares that the New York scenes "betray a strain for effect" (156).

22. Country Life reformers hoped to contain the lingering political passions of populism and post–World War I agrarian radicalism (Fink 25; Neth 105).

23. Other examples are "the colour of the broomsedge was overrunning the desolate hidden field of her life" (64) and "the area of feeling within her soul was parched and blackened, like an abandoned field after the broomsedge is destroyed" (173). Glasgow notes in her 1933 preface: "The country is as familiar to me as if the landscape unrolled both within and without" (viii).

24. For another view, see Godbold, *Ellen Glasgow and the Woman Within*, 148–49. E. Stanly Godbold asserts that "Ellen Glasgow played to the hilt the agrarian image as a symbol of human nature five years before the Southern literary mythmakers took their stand in Nashville" (148). See also Caldwell, "Ellen Glasgow and the Southern Agrarians."

## Chapter Three

1. Warren French asserts that Mitchell's work was an "answer, though indirect," to *The Grapes of Wrath* (*Companion* 136). But Susan Shillinglaw notes that "records indicate that she did indeed write in response to Steinbeck's text" (150 n. 14).

2. All page references to *The Grapes of Wrath* are from John Steinbeck, *The Grapes of Wrath*, introduction by Robert DeMott (New York: Penguin, 1992).

3. All page references to *Of Human Kindness* are from Ruth Comfort Mitchell, *Of Human Kindness* (New York: D. Appleton-Century Co., 1940).

4. Though McWilliams was merciless in his review, *Of Human Kindness* was re-

viewed favorably in the *New York Herald Tribune* (Bell) and the *Christian Century*. In his analysis, McWilliams declares that Mitchell was a direct supporter of the Associated Farmers, the big growers of California ("Glory").

5. Various scholars date the Farmall's release anywhere from 1923 to 1925 (R. C. Williams 86 n. 3). In 1941 Carey McWilliams observed that the tractor "has been the real spearhead of the industrial revolution in agriculture" ("Farms" 413).

6. The idea that one could be tractored out was noted at the time. See R. C. Smith, "New Conditions," 813. In 1941 Walter John Marx declared that "unquestionably the most important cause of tenant displacement is mechanization, in particular, the tractor" (62). Early in *The Grapes of Wrath* a truck driver asks Tom Joad: "A forty-acre cropper and he ain't been dusted out and he ain't been tractored out?" (12). Chapter 5 details the "tractoring out" of a representative tenant farmer (47–53). Of Oklahomans, *Of Human Kindness* notes: "They've been dusted out of their farms, tractored out" (68).

7. Frank J. Taylor, an apologist for the Associated Farmers who investigated flooding in Madera County in the winter of 1937–38, tells a similar story. Responding to Steinbeck's novel, Taylor describes the rescue of eight hundred migrants by health officer Dr. Lee A. Stone (235). During the same winter, a horrified Steinbeck investigated the plight of five thousand migrant families stranded by floods near Visalia (Benson 369). His experience influenced the final scenes of *The Grapes of Wrath* (DeMott, Introduction to *Working* 134).

8. The American Farm Bureau "expressed from the beginning the outlook of the most conservative and prosperous farmers. . . . At the time of its founding, Henry C. Wallace, editor of *Wallace's Farmer* and later secretary of agriculture under Warren Harding, delivered an influential address in which he urged: 'This federation must get to work at once on a real business program if it is to justify its existence. *That doesn't mean turning the work over to committees of farmers either. Every line of work must be in charge of experts*. . . . This federation must not degenerate into an educational or social institution. It must be made the most powerful business institution in the country'" (Hofstadter 127).

9. For the texts of both songs, see *Heath Anthology of American Literature*, 2692–94.

## Chapter Four

1. In its 3 May 1993 issue *Agweek* noted that "people of all walks of life [were] drawn both by the David-vs.-Goliath struggle and by Chavez's self-effacing style" to Chavez's movement ("Cesar Chavez . . . Dies").

2. Five days after Chavez's death, on the day of his funeral, the *Los Angeles Times* reported a "disheartening erosion of many of the gains achieved in the last 25 years. . . . Advocate groups such as California Rural Legal Assistance say they are once again pressing cases of slavery-style treatment of workers by employers" (Arax and Warren).

3. For the uniqueness of Teatro Campesino and its actos, see Ludwig, Introduction, 12, and Cárdenas de Dwyer, "Development of Chicano Drama," 160. Teatro Campesino performed regularly on the pilgrimage to Sacramento, March 1966 (Dunne 133). Both Chavez and Valdez had close ties to Delano: Chavez met his wife there and lived there, and Valdez was born there.

4. Mark Day, in his *Forty Acres: Cesar Chavez and the Farm Workers* (1971), also notes this polarization: "Although some of the workers, including Cesar Chavez and his family, live on the east side, Delano is separated physically as well as racially by the Southern Pacific Railroad tracks. The town's Anglo population lives on the east side, with the Mexican-Americans, Filipinos, and blacks on the west side" (125). Day points out that "Chavez . . . claimed that racism is common in California agriculture" (94).

5. All page references to both actos are from Luis Valdez, *Luis Valdez—Early Works: Actos, Bernabé and Pensamiento Serpentino* (Houston, Tex.: Arte Público, 1990). These actos represent different periods in Teatro Campesino's development. The early acto, *Dos Caras*, reflects the theater's almost total concern with farmworkers; *Vietnam Campesino*, though very much about farmworkers, reflects the theater's movement away from farmworker concerns toward other issues (Yarbro-Bejarano 176–77).

6. Actual working conditions were not hospitable. The trucks were notoriously unsafe, working under the vines meant working amid "sulphur and other pesticides" (Levy 75), and the work was anything but cool and pleasant as the Patroncito suggests: "The workers hunch under the vines like ducks. There is no air, making the intense heat all but unbearable. Gnats and bugs swarm out from under the leaves. Some workers wear face masks; others, handkerchiefs knotted around their heads to catch the sweat" (Dunne 16).

7. See p. 102:

> "BUTT: . . . Remember what we used to do during the grape strike?
> GENERAL: (*Begins to scratch* BUTT'*s back.*) Sure. I used to buy all your scab grapes and ship 'em to Nam . . . give 'em to the boys."

See Reed, "Pentagon Faces a Suit"; "Pentagon Accused"; and Bigart, "Goldberg Calls Election a Referendum."

8. Agent Orange is a herbicide. Its use in Vietnam caused a national controversy in 1979, when "thousands of veterans, armed with the knowledge that their chronic ailments may be related to herbicides, were insistently knocking on the Veterans Administration's door demanding care and compensation" (Uhl and Ensign 210). For photographs of herbicide spraying in Vietnam and the damage it caused, see Uhl and Ensign, *GI Guinea Pigs*, photos between pp. 144 and 145; for maps of Vietnam detailing areas where Agent Orange was sprayed, see pp. 236–39.

9. I thank Jack Temple Kirby for alerting me to this point.

10. William Grami, a leader of the Teamsters Union at Delano, "always referred to the NFWA as the 'Vietcong'" (Dunne 158). Butt likens the phony union to the Saigon government. Later Don Coyote appears as "the democratic, pro-U.S. South Vietnamese president, Ding Dong Diem," who repeats dutifully everything the General tells him to say (114). According to the *New York Times*, Chavez claimed that the Teamsters signed "'sweetheart contracts'" with growers (Lindsey 29). *El Malcriado* depicted this situation in its cartoons: in one, growers and teamsters are in bed together; in another, a huge Teamster figure cradles a grower, crooning, "I'll protect you, sweetheart," while nearby campesinos cry, "They sold us out again!" (in Levy, between pp. 148 and 149).

11. For a description of sharecropping, see McGee and Boone, *Black Rural Land-*

*owner*, 6. Blacks represented 14.3 percent of all farmers in 1920 and 2.3 percent in 1978 (Banks 2).

12. For the role that industrialization and racism have played in the disappearance of the black farmer, see Browning, *Decline of Black Farming*, and Schweninger, "Vanishing Breed," esp. 54–55. McGee and Boone point out: "There is an unequal distribution of resources across economic classes of farms, and the inequality is greater when race is taken into consideration" (50).

13. Of sugar farms surveyed in 1936, 83 percent were worked by resident laborers, with 3 percent worked by sharecroppers, almost all the latter on small farms (Hoffsommer 12).

14. In this context, Creoles are of African and French ancestry. They stand between Cajuns and blacks in Louisiana's social and economic hierarchy (Doyle 79–80).

15. All page references to *A Gathering of Old Men* are from Ernest Gaines, *A Gathering of Old Men* (New York: Vintage, 1992). The tension between Cajuns and plantation owners can be seen when Tee Jack, the Cajun bar owner, describes Jack Marshall, Candy's uncle: "Though he came here to drink, sometimes even bought a round, he would let you know he was not of your crowd" (158). Gil angrily tells Candy Marshall: "You never liked any of us. Looking at us as if we're a breed below you" (122).

16. Gaines alternates perspective with each chapter, using fifteen different narrators—black, white, and Cajun. For his use of alternating narrators, see Gaudet, "Gaines' Fifteen Narrators." For his use of names, see Griffin, "Calling, Naming."

17. That the motor is still running we note on pp. 5–6, 52, 55–56, 64–65.

18. Comparisons of tractor power and costs with horses' were common (R. C. Williams 21). See "Cost Data," 805–7. In 1977 Wendell Berry pointed out that many people believe there are "places in agriculture and forestry that can be competently and economically filled by horses or mules"; their assertion, however, drew a "full-scale attack from the Department of Agriculture" (*Unsettling* 202). Gene Lodgson describes a present-day bustling "horse-drawn economy" (141–53).

19. The novel is based on a story written solely from Dimes's perspective. See Gaines, "Revenge."

## Chapter Five

1. According to Hightower and DeMarco, extension's "mandate is to go among the people of rural America to help them 'identify and solve their farm, home, and community problems through use of research findings of the Department of Agriculture and the State Land Grant Colleges' " (148–49).

2. Many see Iowa as the epitome of American family farming—worry over the family farm's imminent disappearance during the 1980s Farm Crisis centered on reports from hard-hit Iowa, the heart of the Heartland. In studies of farm families, the state's medium-sized family farm base has often been contrasted with California agriculture's emphasis on big family farms; see Friedberger, *Farm Families and Change*, 14. For an argument on how rural areas are colonies of urban centers, see Berry, "Decolonizing Rural America."

3. All page references to *A Thousand Acres* are from Jane Smiley, *A Thousand Acres* (New York: Fawcett Columbine, 1992).

4. For another discussion of similar issues, see Carden, "Remembering/Engendering the Heartland."

5. All page references to *Moo* are from Jane Smiley, *Moo* (New York: Ivy Books, 1998).

6. In the Cooks' new hog confinement system, which is described as a "closed loop," hogs are literally confined on a grid—a "slatted steel floor" (168, 253).

7. Archer-Daniels-Midland, "the world's largest processor of agricultural products," airs its public service commercials on Sunday morning because it believes that the political talk show audience consists of "the movers and shakers in the realms of investment and banking along with politics" (Kahn vii, 10). ADM targets this audience because these people "need to hear" about soil conservation and hunger (Kahn 10). Of course, keeping the corporation's name before such an audience might move investment its way.

8. An epidemic of foot-and-mouth disease swept Great Britain and the Continent in 2000–2001, resulting in the slaughter of hundreds of thousands of animals. See Hoge, "Foot-and-Mouth"; Max, "Foot-and-Mouth."

9. Neill Schaller, associate director of the Institute for Alternative Agriculture, Greenbelt, Maryland, defines low-input sustainable agriculture (LISA) as "a form of agriculture that will not only be productive and profitable for generations to come but will also conserve resources, protect the environment, and enhance the health and safety of the citizenry. Other versions of the same ideal or different paths to it, are known by names such as organic, regenerative, biological, ecological, biodynamic, sustainable, low-input, reduced-input, and alternative agriculture" (22).

10. Beus and Dunlap's work has been a model for other researchers exploring similar or related topics. See Walter and Reisner, "Midwestern Land-Grant University Scientists' Definitions of Sustainable Agriculture."

11. All page references to *Remembering* are from Wendell Berry, *Remembering* (San Francisco: North Point Press, 1988).

12. Scholars have long defined Berry's work as georgic. See Altherr, " 'The Country We Have Married.' " Few book-length studies devoted to Berry's work have appeared. For examples, see Angyal, *Wendell Berry*, and Merchant, *Wendell Berry*.

13. All references to "The Farm" are from Wendell Berry, "The Farm," in *A Timbered Choir* (Washington, D.C.: Counterpoint, 1998), 135–48. Line numbers to the *Georgics* are from book one of Virgil, *The Georgics*, trans. L. P. Wilkinson (London: Penguin Books, 1982).

## Postscript

1. I have changed the names of people mentioned in this section.

# WORKS CITED


Abbott, Richard H. "The Agricultural Press Views the Yeoman: 1819–1859." *Agricultural History* 42, no. 1 (January 1968): 35–48.

Adams, Edward F. *The Modern Farmer in His Business Relations*. San Francisco: N. J. Stone Co., 1899.

Alpers, Paul. *What Is Pastoral?* Chicago: University of Chicago Press, 1996.

Altherr, Thomas L. "'The Country We Have Married': Wendell Berry and the Georgic Tradition of Agriculture." *Southern Studies* 1, no. 2 (Summer 1990): 105–15.

Anderson, Frederick Irving. *The Farmer of To-Morrow*. New York: Macmillan, 1914.

Angyal, Andrew J. *Wendell Berry*. Boston: Twayne, 1995.

"An Appeal to the People." *Visalia Weekly Delta*, 7 May 1880.

Arax, Mark, and Jennifer Warren. "Chavez's Gains Are Slipping Away." *Los Angeles Times*, Orange County ed., 29 April 1993, 13–14.

Atkeson, Mary Meek. *The Woman on the Farm*. New York: Century Co., 1924.

Bagby, Beth. "El Teatro Campesino: Interviews with Luis Valdez." *Tulane Drama Review* 11, no. 4 (Summer 1967): 70–80.

Bailey, L. H. *The Country-Life Movement in the United States*. New York: Macmillan, 1911.

Baker, Ray Stannard. "The Movement of Wheat." *McClure's Magazine* 14 (December 1899): 124–37.

Banks, Vera J. *Black Farmers and Their Farms*. Rural Development Research Report Number 59. Washington, D.C.: U.S. Department of Agriculture, 1986.

Barriga, Joan B. "Lark on a Barbed-Wire Fence: Ruth Comfort Mitchell's 'Of Human Kindness.'" *Steinbeck Newsletter* (Summer 1989): 3–4.

Bartley, Numan V. "The Southern Enclosure Movement." Review of *Rural Worlds Lost: The American South, 1920–1960*, by Jack Temple Kirby. *Georgia Historical Quarterly* 71, no. 3 (Fall 1987): 438–50.

Bassnett-McGuire, Susan. "El Teatro Campesino: From Actos to Mitos." *Theatre Quarterly* 9, no. 34 (Summer 1979): 18–21.

Bates, J. Leonard. *The United States, 1898–1928: Progressivism and a Society in Transition*. New York: McGraw-Hill, 1976.

Baum, Rosalie Murphy. "The Burden of Myth: The Role of the Farmer in American Literature." *North Dakota Quarterly* 53, no. 4 (Fall 1985): 4–24.

Beall, Robert T. "Rural Electrification." In *Farmers in a Changing World: The Yearbook of Agriculture, 1940*, edited by Gove Hambidge, 790–809. Washington, D.C.: Government Printing Office, 1940.

Bell, Lisle. Review of *Of Human Kindness*, by Ruth Comfort Mitchell. *New York Herald Tribune*, 5 May 1940, 19.

Benson, Jackson J. *The True Adventures of John Steinbeck, Writer*. New York: Viking, 1984.

Benton, Alva H. "Large Land Holdings in North Dakota." *Journal of Land and Public Utility Economics* 1, no. 4 (October 1925): 405–13.

Berry, Wendell. "Decolonizing Rural America." *Audubon*, March–April 1993, 100–105.

———. "The Farm." *A Timbered Choir*. Washington, D.C.: Counterpoint, 1998: 135–48.

———. Preface. *A Timbered Choir*. Washington, D.C.: Counterpoint, 1998: xvii–xviii.

———. *Remembering*. San Francisco: North Point, 1988.

———. *Standing by Words*. San Francisco: North Point, 1983.

———. *The Unsettling of America*. San Francisco: Sierra Club Books, 1986.

———. "Writer and Region." *Hudson Review* 40, no. 1 (Spring 1987): 15–30.

Beus, Curtis E., and Riley E. Dunlap. "Conventional versus Alternative Agriculture: The Paradigmatic Roots of the Debate." *Rural Sociology* 55, no. 4 (1990): 590–616.

Bigart, Homer. "Goldberg Calls Election a Referendum on Nixon Administration." *New York Times*, 1 October 1970, 35.

Bigelow, Poultney. "The Bonanza Farms of the West." *Atlantic Monthly* 45 (January 1880): 33–44.

Blair, William M. "Agriculture Nominee: Earl Lauer Butz." *New York Times*, 12 November 1971, 33.

Blank, Steven C. *The End of Agriculture in the American Portfolio*. Westport, Conn.: Quorum Books, 1998.

Bond, Tonette L. "Pastoral Transformations in *Barren Ground*." *Mississippi Quarterly* 32 (Fall 1979): 565–76.

Bowers, William L. *The Country Life Movement in America, 1900–1920*. Port Washington, N.Y.: Kennikat Press, 1974.

"Boycott Unit Begins Drive on Pesticides." *New York Times*, 27 January 1969, 9.

Boynton, H. W. "Back to the Soil." Review of *Barren Ground*, by Ellen Glasgow. *Providence Journal*, 12 April 1925, 7, sec. G. (Reprint, in *Ellen Glasgow: The Contemporary Reviews*, edited by Dorothy M. Scura, 249. Cambridge: Cambridge University Press, 1992.)

Briggs, Harold E. "Early Bonanza Farming in the Red River Valley of the North." *Agricultural History* 6, no. 1 (January 1932): 26–37.

Brock, H. I. "Southern Romance Is Dead." Review of *Barren Ground*, by Ellen Glasgow. *New York Times Book Review*, 12 April 1925, 2, sec. 3. (Reprint, in *Ellen Glasgow: The Contemporary Reviews*, edited by Dorothy M. Scura, 246–49. Cambridge: Cambridge University Press, 1992.)

Brown, Edward G. "The Teatro Campesino's Vietnam Trilogy." *Minority Voices* 4, no. 1 (1980): 29–38.

Brown, T. "What a Farmer Really Looks Like." Cartoon. *Country Gentleman*, 5 November 1921, 4.

Browning, Pamela. *The Decline of Black Farming in America*. U.S. Commission on Civil Rights Report. Washington, D.C.: Government Printing Office, 1982.

Buell, Lawrence. *The Environmental Imagination*. Cambridge: Harvard University Press, 1995.

Burkett, Charles W. "What the Farm Home Needs." *Good Housekeeping* 48, no. 2 (February 1909): 148–50.

"The Business of a Wheat Farm." *Cultivator and Country Gentleman*, 11 November 1897, 896.

Butterfield, Kenyon L. *Chapters in Rural Progress*. Chicago: University of Chicago Press, 1907.

Caldwell, Ellen M. "Ellen Glasgow and the Southern Agrarians." *American Literature* 56, no. 2 (May 1984): 203–13.

*California Guide Book*. Pamphlet. San Francisco: Pacific Coast Land Bureau, n.d. (post–1882).

Carden, Mary Paniccia. "Creative Fertility and the National Romance in Willa Cather's *O Pioneers!* and *My Antonia*." *Modern Fiction Studies* 45, no. 2 (Summer 1999): 275–302.

———. "Remembering/Engendering the Heartland: Sexed Language, Embodied Space, and America's Foundational Fictions in Jane Smiley's *A Thousand Acres*." *Frontiers* 18, no. 2 (1997): 181–202.

Cárdenas de Dwyer, Carlota. "The Development of Chicano Drama and Luis Valdez's *Actos*." In *Modern Chicano Writers*, edited by Joseph Sommers and Tomás Ybarra-Frausto, 160–66. Englewood Cliffs, N.J.: Prentice-Hall, 1979.

Cargill, Oscar. Afterword to *The Octopus*, by Frank Norris. New York: New American Library, 1981.

Carlson, Ron. "King Lear in Zebulon County." Review of *A Thousand Acres*, by Jane Smiley. *New York Times Book Review*, 3 November 1991, 12.

Casson, Herbert N. "The New American Farmer." *Review of Reviews* 37 (May 1908): 598–602.

Cather, Willa. "Nebraska: The End of the First Cycle." *Nation*, 5 September 1923, 236–38.

———. *O Pioneers!* Edited by Susan J. Rosowski and Charles W. Mignon. Lincoln: University of Nebraska Press, 1992.

Cato (the Censor). *On Agriculture*. In *Cato and Varro on Agriculture*, translated by William Davis Hooper and edited by G. P. Goold, 2–157. Cambridge: Harvard University Press, 1993.

"Cesar Chavez." Obituary. *California Farmer*, June 1993, 40.

"Cesar Chavez, the Farm Worker's Champion, Dies." *Agweek*, 3 May 1993, 39.

Chapin, A. B. Cartoon. *Country Gentleman*, 2 July 1921, 7.

"Chavez Scores Grape Growers on Pesticide Use." *New York Times*, 30 September 1969, 19.

Cherny, Robert W. "Nebraska, 1883–1925: Cather's Version and History's." In *Willa Cather: Family, Community, and History*, edited by John J. Murphy, 229–51. Provo, Utah: Brigham Young University Humanities Publications Center, 1990.

———. "Willa Cather and the Populists." *Great Plains Quarterly* 3 (Fall 1983): 206–18.

Clanton, Gene. "Populism, Progressivism, and Equality: The Kansas Paradigm." *Agricultural History* 51, no. 3 (1977): 559–81.

Cochran, Caroline. Letter. *Washington Post*, 4 June 1994, A17.

Coffin, C. C. "Dakota Wheat Fields." *Harper's New Monthly Magazine* 60 (March 1880): 529–35.

Collins, T. Byard. *The New Agriculture*. New York: Munn and Co., 1906.

"A Collision." *San Francisco Chronicle*, 12 May 1880.
Comstock, Gary, ed. *Is There a Moral Obligation to Save the Family Farm?* Ames: Iowa State University Press, 1987.
"Cost Data for Farm Products." *U.S. Department of Agriculture Yearbook*, 1921. Washington, D.C.: Government Printing Office, 1922.
Crawford, Lester M. "BSE: A Veterinary History." In *The Mad Cow Crisis: Health and the Public Good*, edited by Scott C. Ratzan, 9–14. New York: New York University Press, 1998.
Crawford, Lewis F. *History of North Dakota*. Vol. 1. Chicago: American Historical Society, 1931.
Crawford, Robert Platt. *These Fifty Years: A History of the College of Agriculture of the University of Nebraska*. Lincoln: University of Nebraska Press, 1925.
Danbom, David B. *The Resisted Revolution: Urban America and the Industrialization of Agriculture, 1900–1930*. Ames: Iowa State University Press, 1979.
Daniel, Cletus E. *Bitter Harvest: A History of California Farmworkers, 1870–1941*. Ithaca: Cornell University Press, 1981.
Davenport, Eugene. "Scientific Farming." In *Country Life*, edited by J. P. Lichtenberger, 45–50. Philadelphia: American Academy of Political and Social Science, 1912.
Davison, Richard Allan. *The Merrill Studies in The Octopus*. Columbus, Ohio: Charles E. Merrill Publishing Co., 1969.
Day, Mark. *Forty Acres: Cesar Chavez and the Farm Workers*. Introduction by Cesar Chavez. New York: Praeger Publishers, 1971.
DeMott, Robert. Introduction to *The Grapes of Wrath*, by John Steinbeck, vii–xliv. New York: Penguin, 1992.
———. Introduction to *Working Days: The Journals of The Grapes of Wrath, 1938–1941*, edited by DeMott, xxi–lvii. New York: Viking, 1989.
Den Tandt, Christophe. *The Urban Sublime in American Literary Naturalism*. Urbana: University of Illinois Press, 1998.
Dillingham, William B. *Frank Norris: Instinct and Art*. Lincoln: University of Nebraska Press, 1969.
Dillon, Charles. "Education and Discontent." *Outlook*, 10 April 1909, 829–32.
Doyle, Mary Ellen. "Ernest Gaines' Materials: Place, People, Author." *MELUS* 15, no. 3 (Fall 1988): 75–93.
Drache, Hiram M. *The Day of the Bonanza*. Fargo: North Dakota Institute for Regional Studies, 1964.
Duncan, Charles. "'If Your View Be Large Enough': Narrative Growth in *The Octopus*." *American Literary Realism* 25, no. 2 (Winter 1993): 56–66.
Dunne, John Gregory. *Delano: The Story of the California Grape Strike*. New York: Farrar, Straus and Giroux, 1967.
"Earl Butz versus Wendell Berry." *Co-Evolution Quarterly* 17 (Spring 1978): 50–59.
Easterbrook, Gregg. "Making Sense of Agriculture: A Revisionist Look at Farm Policy." In *Is There a Moral Obligation to Save the Family Farm?*, edited by Gary Comstock, 3–30. Ames: Iowa State University Press, 1987.
Eichenwald, Kurt. "Archer Daniels Agrees to Big Fine for Price Fixing." *New York Times*, 15 October 1996, A1, D7.
Ellenberg, George B. "Debating Farm Power: Draft Animals, Tractors, and the

United States Department of Agriculture." *Agricultural History* 74, no. 2 (Spring 2000): 545–68.

Ellis, Reuben J. "'A Little Turn through the Country': Presley's Bicycle Ride in Frank Norris's *The Octopus*." *Journal of American Culture* 17, no. 3 (Fall 1994): 17–22.

Ellsworth, Clayton S. "Theodore Roosevelt's Country Life Commission." *Agricultural History* 34 (October 1960): 155–72.

"EPA Wants Tighter Animal Feedlot Rules." *Washington Post*, 16 December 2000, A15.

Esbjornson, Carl D. "*Remembering* and Home Defense." In *Wendell Berry*, edited by Paul Merchant, 155–70. Lewiston, Idaho: Confluence Press, 1991.

Ettin, Andrew V. *Literature and the Pastoral*. New Haven: Yale University Press, 1984.

"Fact Sheet: What Is the American Farmland Trust?" 13 March 2001. 30 March 2001. http://www.farmland.org/files/steward/factsheet.htm.

"Family's Dairy Farm Finally Swallowed Up by Suburbia." *Tribune* (Scranton, Pa.), 8 August 1994, B8.

Farmall. Advertisement. *Country Gentleman*, January 1939.

———. Advertisement. *Country Gentleman*, April 1939.

———. Advertisement. *Hoard's Dairyman*, 10 January 1939.

"Farmers' Four Peace Points." *Stockton Daily Evening Record*, 8 December 1939, 1, 21.

Feder, Barnaby J. "For Amber Waves of Data." *New York Times*, 4 May 1998, D1, D4.

———. "Sowing Preservation." *New York Times*, 20 March 1997, D1, D19.

Feingold, Richard. *Nature and Society: Later Eighteenth-Century Uses of the Pastoral and Georgic*. New Brunswick: Rutgers University Press, 1978.

Feldman, Shelley, and Rick Welsh. "Feminist Knowledge Claims, Local Knowledge, and Gender Divisions of Agricultural Labor: Constructing a Successor Science." *Rural Sociology* 60, no. 1 (1995): 25–43.

Fetherston, David. *Farm Tractor Advertising in America, 1900–1960*. Osceola, Wis.: Motorbooks, 1996.

Fiedler, Leslie. *Love and Death in the American Novel*. New York: Anchor Books, 1992.

Fink, Deborah. *Agrarian Women: Wives and Mothers in Rural Nebraska, 1880–1940*. Chapel Hill: University of North Carolina Press, 1992.

Fisher, Ann H. Review of *A Thousand Acres*, by Jane Smiley. *Library Journal*, 1 October 1991, 142.

Fisher, Philip. "Democratic Social Space: Whitman, Melville, and the Promise of American Transparency." In *The New American Studies*, edited by Philip Fisher, 70–111. Berkeley: University of California Press, 1991.

Fite, Gilbert C. *American Farmers: The New Minority*. Bloomington: Indiana University Press, 1981.

———. *Cotton Fields No More: Southern Agriculture, 1865–1980*. Lexington: University Press of Kentucky, 1984.

Flanagan, John T. "The Middle Western Farm Novel." *Minnesota History* 23, no. 2 (June 1942): 113–25.

Foght, Harold W. "The Country School." In *Country Life*, edited by J. P. Lichtenberger, 149–57. Philadelphia: American Academy of Political and Social Science, 1912.

French, Warren. *Frank Norris*. New York: Twayne, 1962.
———, ed. *A Companion to The Grapes of Wrath*. New York: Viking, 1963.
Friedberger, Mark. *Farm Families and Change in Twentieth-Century America*. Lexington: University Press of Kentucky, 1988.
"Fruit Concern Sells 14-Million Property Planted in Grapes." *New York Times*, 20 April 1969, 56.
Fuller, Wayne. "Making Better Farmers: The Study of Agriculture in Midwestern Country Schools, 1900–1923." *Agricultural History* 60, no. 2 (Spring 1986): 154–68.
———. "The Rural Roots of the Progressive Leaders." *Agricultural History* 42, no. 1 (1968): 1–13.
Gaines, Ernest. *A Gathering of Old Men*. New York: Vintage Books, 1992.
———. "The Revenge of the Old Men." *Callaloo* 1, no. 3 (May 1978): 5–21.
Galpin, Charles Josiah. *Rural Life*. New York: Century Co., 1920.
———. *Rural Social Problems*. New York: Century Co., 1924.
Gardner, Gary. *Shrinking Fields: Cropland Loss in a World of Eight Billion*. Washington, D.C.: Worldwatch Institute, 1996.
Gates, Paul. *Agriculture and the Civil War*. New York: Alfred A. Knopf, 1965.
Gaudet, Marcia. "Gaines' Fifteen Narrators: Narrative Style and Storytelling Technique in *A Gathering of Old Men*." *Louisiana Folklore Miscellany* 6, no. 3 (1990): 15–22.
Gaudet, Marcia, and Carl Wooton. *Porch Talk with Ernest Gaines*. Baton Rouge: Louisiana State University Press, 1990.
Gentilcore, Roxanne M. "American Georgic: Vergil in the Literature of the Colonial South." *Classical and Modern Literature* 13, no. 3 (1993): 257–70.
Gerber, Philip L. *Willa Cather*. Boston: Twayne, 1975.
Gilman, Charlotte Perkins. "That Rural Home Inquiry." *Good Housekeeping* 48, no. 1 (January 1909): 120–22.
Glasgow, Ellen. *Barren Ground*. San Diego: Harcourt Brace Jovanovich, 1985.
———. *A Certain Measure*. New York: Harcourt, Brace, 1943.
———. Preface. *Barren Ground*. San Diego: Harcourt Brace Jovanovich, 1985: vii–ix.
———. *The Woman Within*. New York: Harcourt, Brace, 1954.
Glotfelty, Cheryll. Introduction to *The Ecocriticism Reader: Landmarks in Literary Ecology*, edited by Glotfelty and Harold Fromm, xv–xxxvii. Athens: University of Georgia Press, 1996.
Godbold, E. Stanly, Jr. *Ellen Glasgow and the Woman Within*. Baton Rouge: Louisiana State University Press, 1972.
Gollner, Philipp M. "Thousands in California Say Goodbye to Chavez." *New York Times*, 30 April 1993, 13.
Goodwyn, Lawrence. *Democratic Promise: The Populist Moment in America*. New York: Oxford University Press, 1976.
Graham, Don. *The Fiction of Frank Norris: The Aesthetic Context*. Columbia: University of Missouri Press, 1978.
Grant, Gary. Black Farmers and Agriculturalists Association. Home page. April 1998. 22 January 2000. "NC A&T Event Gives Voice to Black Farmers." http://www.Ncat.edu/~soa/news/apr98/blackfarmers.html.

Grantham, Dewey W., Jr. "The Progressive Era and the Reform Tradition." In *Progressivism: The Critical Issues*, edited by David M. Kennedy, 109–21. Boston: Little, Brown, 1971.
Green, Elizabeth Lay. Review of *Barren Ground*, by Ellen Glasgow. *Reviewer* 5, no. 4 (October 1925): 118–19.
Gregory, James N. *American Exodus: The Dust Bowl Migration and Okie Culture in California*. New York: Oxford University Press, 1989.
Griffin, Joseph. "Calling, Naming, and Coming of Age in Ernest Gaines's 'A Gathering of Old Men.'" *Names* 40, no. 2 (June 1992): 89–97.
Gustafson, Neil. "Getting Back to Cather's Text: The Shared Dream in *O Pioneers!*" *Western American Literature* 30, no. 2 (Summer 1995): 151–62.
Ham, William T. "Farm Labor in an Era of Change." In *Farmers in a Changing World: Yearbook of Agriculture, 1940*, edited by Gove Hambidge, 907–21. Washington, D.C.: Government Printing Office, 1940.
"Hamburger Plant Closing, Officials Beef Up Recall." *Tribune* (Scranton, Pa.), 22 August 1997, A1.
Harger, Charles Moreau. "The New Life on the Farm." *Outlook*, 16 April 1910, 841–44.
Harrison, Elizabeth Jane. *Female Pastoral: Women Writers Re-Visioning the American South*. Knoxville: University of Tennessee Press, 1991.
Harrop, John, and Jorge Huerta. "The Agitprop Pilgrimage of Luis Valdez and El Teatro Campesino." *Theatre Quarterly* 5, no. 17 (March–May 1975): 30–39.
Harwood, W. S. *The New Earth: A Recital of the Triumphs of Modern Agriculture in America*. New York: Macmillan, 1906.
Hay, James, Jr. "The Man from the City." In *Early Stories from the Land: Short-Story Fiction from American Rural Magazines, 1900–1925*, edited by Robert G. Hays, 233–49. Ames: Iowa State University Press, 1995.
*The Heath Anthology of American Literature*. Edited by Paul Lauter. 2d ed. Vol. 1. Lexington, Mass.: D. C. Heath, 1994.
Heinzelman, Kurt. "Roman Georgic in the Georgian Age: A Theory of Romantic Genre." *Texas Studies in Literature and Language* 33, no. 2 (Summer 1991): 182–214.
Henderson, Archibald. "Soil and Soul." Review of *Barren Ground*, by Ellen Glasgow. *Saturday Review*, 18 July 1925, 907. (Reprint, in *Ellen Glasgow: The Contemporary Reviews*, edited by Dorothy M. Scura, 264–65. Cambridge: Cambridge University Press, 1992.)
Hernández, Guillermo E. *Chicano Satire: A Study in Literary Culture*. Austin: University of Texas Press, 1991.
Hesiod. *The Works and Days, Theogony, and the Shield of Herakles*. Translated by Richard Lattimore. Ann Arbor: University of Michigan Press, 1991.
Hicks, John D. *The Populist Revolt*. Lincoln: University of Nebraska Press, 1961.
"Highlights of Agriculture: 1997 and 1992, United States." *1997 Census of Agriculture*. 1997. 19 May 1999. http://www.nass.usda.gov/census/census97/highlights/usasum/us.txt.
Hightower, Jim. *Hard Times, Hard Tomatoes*. Cambridge, Mass.: Schenkman Publishing, 1973.
Hightower, Jim, and Susan DeMarco. "Hard Tomatoes, Hard Times: The Failure

of the Land Grant College Complex." In *Is There a Moral Obligation to Save the Family Farm?*, edited by Gary Comstock, 135–52. Ames: Iowa State University Press, 1987.

Hill, Andrew F., Melanie Desbruslais, Susan Joiner, Katie C. L. Sidle, Ian Gowland, and John Collinge. "The Same Prion Strain Causes vCJD and BSE." *Nature*, 2 October 1997, 448–50.

Hillerton, J. Eric. "Bovine Spongiform Encephalopathy: Current Status and Possible Impacts." *Journal of Dairy Science* 81, no. 11 (November 1998): 3042–48.

Hochman, Barbara. *The Art of Frank Norris, Storyteller*. Columbia: University of Missouri Press, 1988.

Hoffsommer, Harold C. *The Sugar Cane Farm: A Social Study of Labor and Tenancy*. Louisiana Bulletin Number 320. Baton Rouge: Louisiana State University and Agricultural and Mechanical College Agricultural Experiment Station, 1940.

Hofstadter, Richard. *The Age of Reform*. New York: Vintage, 1955.

Hoge, Warren. "Foot-and-Mouth out of Control." *Scranton Times*, 24 March 2001, 1–2.

Holmes, Roy Hinman. "The Passing of the Farmer." *Atlantic Monthly* 110, no. 4 (October 1912): 517–23.

Holt, Marilyn Irvin. "From Better Babies to 4-H: A Look at Rural America, 1900–1930." *Prologue* 24, no. 3 (1992): 245–55.

Horwitz, Howard. *By the Law of Nature: Form and Value in Nineteenth-Century America*. New York: Oxford University Press, 1991.

Howard, Ann. "Framework for Work Change." In *The Changing Nature of Work*, edited by Howard, 3–44. San Francisco: Jossey-Bass Publishers, 1995.

Hoy, Suellen. *Chasing Dirt: The American Pursuit of Cleanliness*. New York: Oxford University Press, 1995.

Huerta, Jorge A. "Labor Theatre, Street Theatre and Community Theatre in the Barrios, 1965–1983." In *Hispanic Theatre in the United States*, edited by Nicolás Kanellos, 62–70. Houston: Arte Público Press, 1984.

Jackson, Wes. *Becoming Native to This Place*. Washington, D.C.: Counterpoint, 1996.

Jefferson, Thomas. *Notes on the State of Virginia*. In *The Portable Thomas Jefferson*, edited by Merrill D. Peterson, 23–232. New York: Penguin, 1987.

———. "To Benjamin Austin." 9 January 1816. Letter. In *The Portable Thomas Jefferson*, edited by Merrill D. Peterson, 547–50. New York: Penguin, 1987.

Jellison, Katherine. *Entitled to Power: Farm Women and Technology, 1913–1963*. Chapel Hill: University of North Carolina Press, 1993.

Jewett, Sarah Orne. *The Country of the Pointed Firs and Other Stories*. Preface by Willa Cather. New York: Anchor Books, 1989.

Johnson, Bekki T. Letter. *Washington Post*, 4 June 1994, A17.

Johnstone, Paul H. "Old Ideals versus New Ideas in Farm Life." In *Farmers in a Changing World: The Yearbook of Agriculture, 1940*, edited by Gove Hambidge, 111–70. Washington, D.C.: Government Printing Office, 1940.

"Judge Approves Settlement for Black Farmers." *New York Times*, 15 April 1999, A29.

"Julio Gallo." Obituary. *California Farmer*, June 1993, 40.

Kahn, E. J., Jr. *Supermarketer to the World*. New York: Warner Books, 1991.

Kappel, Tim. "Trampling Out the Vineyards: Kern County's Ban on *The Grapes of Wrath*." *California History* 61, no. 3 (1982): 210–21.

Kifer, R. S., B. H. Hurt, and Albert A. Thornburgh. "The Influence of Technical Progress on Agricultural Production." In *Farmers in a Changing World: The Yearbook of Agriculture, 1940*, edited by Gove Hambidge, 509–32. Washington, D.C.: Government Printing Office, 1940.

King, Seth S. "Block Sees Food as U.S. Weapon in Foreign Policy." *New York Times*, 24 December 1980, 1, 13.

Kirby, Jack Temple. *Rural Worlds Lost: The American South, 1920–1960*. Baton Rouge: Louisiana State University Press, 1987.

Kolodny, Annette. *The Lay of the Land: Metaphor as Experience and History in American Life and Letters*. Chapel Hill: University of North Carolina Press, 1975.

Kourilsky, Françiose. "Approaching Quetzalcoatl: The Evolution of El Teatro Campesino." *Performance* 2, no. 1 (Fall 1973): 37–46.

Kraybill, Donald B. *The Riddle of Amish Culture*. Baltimore: Johns Hopkins University Press, 1989.

Kulikoff, Allan. "The Transition to Capitalism in Rural America." *William and Mary Quarterly* 46 (1989): 120–44.

LaFraniere, Sharon. "Agency Fails to Collect Millions in Loans to Wealthy Farm Owners." *Washington Post*, 28 January 1994, A1, A16.

Lamar, Jacob V., Jr. " 'He Couldn't Manage Any More.' " *Time*, 23 December 1985, 26.

Languish, Lydia. "Down on the Farm." Review of *Barren Ground*, by Ellen Glasgow. *John O'London's Weekly*, 25 July 1925, 528. (Reprint, in *Ellen Glasgow: The Contemporary Reviews*, edited by Dorothy M. Scura, 267. Cambridge: Cambridge University Press, 1992.)

Lawrence, D. H. *Studies in Classic American Literature*. London: Heinemann, 1964.

Leckie, Gloria J. " 'They Never Trusted Me to Drive': Farm Girls and the Gender Relations of Agricultural Information Transfer." *Gender, Place and Culture* 3, no. 3 (1996): 309–25.

Lee, Gary. "Farm Herbicides Foul Tap Water for 14 Million." *Washington Post*, 19 October 1994, A3.

Leopold, Aldo. *A Sand County Almanac*. New York: Ballantine, 1991.

Leslie, Marina. "Incest, Incorporation, and *King Lear* in Jane Smiley's *A Thousand Acres*." *College English* 60, no. 1 (January 1998): 31–50.

LeSueur, Meridel. *Ripening*. Edited and introduction by Elaine Hedges. New York: Feminist Press, 1990.

Levy, Jacques E. *Cesar Chavez: Autobiography of La Causa*. New York: W. W. Norton, 1975.

Light, Frank. "Obit Questions." Letter. *California Farmer*, August 1993, 34.

Lindsey, Robert. "Cesar Chavez, 66, Organizer of Union for Migrants, Dies." *New York Times*, 24 April 1993, 1, 29.

Lisca, Peter. *The Wide World of John Steinbeck*. New Brunswick, N.J.: Rutgers University Press, 1958.

Lodgson, Gene. *At Nature's Pace: Farming and the American Dream*. New York: Pantheon, 1994.

Love, Glen A. "Revaluing Nature: Toward an Ecological Criticism." In *The*

*Ecocriticism Reader: Landmarks in Literary Ecology*, edited by Cheryll Glotfelty and Harold Fromm, 225–40. Athens: University of Georgia Press, 1996.

Low, Anthony. *The Georgic Revolution*. Princeton: Princeton University Press, 1985.

Ludwig, Edward W. Introduction to *The Chicanos: Mexican American Voices*, edited by Edward Ludwig and James Santibañez, 1–22. Baltimore: Penguin Books, 1971.

MacGarr, Llewellyn. *The Rural Community*. New York: Macmillan, 1922.

Manley, Robert N. *Centennial History of the University of Nebraska*. Vol. 1. Lincoln: University of Nebraska Press, 1969.

Marcus, Edwin. "What a Farmer Really Looks Like." Cartoon. *Country Gentleman*, 19 November 1921, 7.

Maris, Paul V. "Farm Tenancy." In *Farmers in a Changing World: The Yearbook of Agriculture, 1940*, edited by Gove Hambidge, 887–906. Washington, D.C.: Government Printing Office, 1940.

Marx, Leo. *The Machine in the Garden: Technology and the Pastoral Ideal in America*. New York: Oxford University Press, 1974.

———. "Pastoral Ideals and City Troubles." *Journal of General Education* 20, no. 4 (January 1969): 251–71.

———. "Pastoralism in America." In *Ideology and Classic American Literature*, edited by Sacvan Bercovitch and Myra Jehlen, 36–69. Cambridge: Cambridge University Press, 1986.

Marx, Walter John. *Mechanization and Culture: The Social and Cultural Implications of a Mechanized Society*. St. Louis, Mo.: B. Herder Book Co., 1941.

Mathews, Jessica. "When a Sea Dies." *Washington Post*, 4 October 1994, A17.

Mathieson, Barbara. "The Polluted Quarry: Nature and Body in *A Thousand Acres*." In *Transforming Shakespeare: Contemporary Women's Re-Visions in Literature and Performance*, edited by Marianne Novy, 127–44. New York: St. Martin's Press, 1999.

Matthiessen, Peter. "Cesar Chavez." *New Yorker*, 17 May 1993, 82.

Max, Arthur. "Foot-and-Mouth Gaining in Europe." *Scranton Times*, 22 March 2001, 10.

McConnell, Grant. *The Decline of Agrarian Democracy*. Berkeley: University of California Press, 1959.

McDowell, Frederick P. W. *Ellen Glasgow and the Ironic Art of Fiction*. Madison: University of Wisconsin Press, 1960.

McGee, Leo, and Robert Boone, eds. *The Black Rural Landowner—Endangered Species: Social, Political and Economic Implications*. Westport, Conn.: Greenwood Press, 1979.

McKerns, Joseph P., ed. *Biographical Dictionary of American Journalism*. New York: Greenwood Press, 1989.

McMath, Robert C., Jr. *American Populism: A Social History, 1877–1898*. New York: Noonday Press, 1993.

McMillen, Wheeler. "Making the Farm Fit the Tractor." *Power Farming* 30, no. 1 (January 1921): 9–10.

McWilliams, Carey. *Factories in the Field: The Story of Migratory Farm Labor in California*. Boston: Little, Brown, 1944.

———. "Farms into Factories: Our Agricultural Revolution." *Antioch Review* (Winter 1941): 406–31.

———. “Glory, Glory California.” Review of *Of Human Kindness*, by Ruth Comfort Mitchell. *New Republic*, 22 July 1940, 125.

Merchant, Paul, ed. *Wendell Berry*. Lewiston, Idaho: Confluence Press, 1991.

Meyer, Roy W. *The Middle Western Farm Novel in the Twentieth Century*. Lincoln: University of Nebraska Press, 1965.

Michaels, Walter Benn. *The Gold Standard and the Logic of Naturalism*. Berkeley: University of California Press, 1987.

Mitchell, Ruth Comfort. *Of Human Kindness*. New York: Appleton-Century, 1940.

Morrison, Toni. *The Bluest Eye*. New York: Washington Square Press, 1970.

Motley, Warren. “The Unfinished Self: Willa Cather’s *O Pioneers!* and the Psychic Cost of a Woman’s Success.” *Women’s Studies* 12, no. 2 (1986): 149–65.

Murphy, John J. “A Comprehensive View of Cather’s *O Pioneers!*” In *Critical Essays on Willa Cather*, edited by John J. Murphy, 113–27. Boston: G. K. Hall, 1984.

Murray, Stanley N. “Railroads and the Agricultural Development of the Red River Valley of the North, 1870–1890.” *Agricultural History* 31, no. 4 (October 1957): 57–66.

———. *The Valley Comes of Age*. Fargo: North Dakota Institute for Regional Studies, 1967.

Naughton, James M. “Hardin Out as Farm Chief; President to Keep Agency.” *New York Times*, 12 November 1971, 1, 32.

Neth, Mary. *Preserving the Family Farm: Women, Community, and the Foundations of Agribusiness in the Midwest, 1900–1940*. Baltimore: Johns Hopkins University Press, 1995.

Newswanger, Everett. “Agribusiness Professionals, Farmers Gather to Discuss Future.” *Lancaster Farming*, 24 December 1994, A1, A28.

“Nixon Agriculture Nominee to Quit Four Corporations.” *New York Times*, 15 November 1971, 27.

Norris, Frank. *The Octopus*. New York: New American Library, 1981.

“Noted Authoress Answers Charges of Novel, Defends Migrants.” *Stockton Daily Evening Record*, 8 December 1939, 1, 21.

Opie, John. *The Law of the Land: Two Hundred Years of American Farmland Policy*. Lincoln: University of Nebraska Press, 1994.

Parker, Suzi. “The Vanishing Black Farmer.” *Christian Science Monitor*, 13 July 2000, 1, 3.

“Pentagon Accused by Chavez for Rise in Grape Purchases.” *New York Times*, 29 September 1969, 96.

Perkell, Christine G. *The Poet’s Truth: A Study of the Poet in Virgil’s Georgics*. Berkeley: University of California Press, 1989.

Peterson, Chester, Jr. “Field Mapping Made Easy.” *Furrow*, December 1995, 31–32.

“Pioneers in Mussel Slough.” *Visalia Weekly Delta*, 4 June 1880.

Pizer, Donald. *The Novels of Frank Norris*. Bloomington: Indiana University Press, 1966.

Power, James B. “Grain Farms in the Northwest.” *Cultivator and Country Gentleman* 8 (January 1880): 19–20.

Preston, William L. *Vanishing Landscapes: Land and Life in the Tulare Lake Basin*. Berkeley: University of California Press, 1981.

Quick, Herbert. "The Women on the Farms." *Good Housekeeping* 57 (1913): 426–36.
*Quick Facts from the Census of Agriculture*. Pamphlet. National Agricultural Statistics Service. Washington, D.C.: U.S. Department of Agriculture, 1999.
Randall, John H., III. *The Landscape and the Looking Glass: Willa Cather's Search for Value*. Boston: Houghton Mifflin, 1960.
Raper, Julius Rowan. "*Barren Ground* and the Transition to Southern Modernism." In *Ellen Glasgow: New Perspectives*, edited by Dorothy M. Scura, 146–61. Knoxville: University of Tennessee Press, 1995.
Reed, Roy. "Pentagon Faces a Suit on Grapes." *New York Times*, 27 June 1969, 17.
*Report of the Commission on Country Life*. Introduction by Theodore Roosevelt. New York: Sturgis and Walton, 1911.
Review of *Of Human Kindness*, by Ruth Comfort Mitchell. *Christian Century*, 1 May 1940, 581–82.
Reynolds, Guy. *Willa Cather in Context: Progress, Race, Empire*. New York: St. Martin's Press, 1996.
Ristau, Kevin, and Mark Ritchie. "The Farm Crisis: History and Analysis." *Shmate* 16 (Fall 1986): 10–20.
Roberts, Stephen V. "Charge of Peril in Pesticides Adds Fuel to Coast Grape Strike." *New York Times*, 16 March 1969, 46.
———. "2 Big Issues Snag Grape Union Talk." *New York Times*, 17 July 1969, 28.
Robinson, W. W. *Land in California*. Berkeley: University of California Press, 1948.
Rohn, Ray. "What a Farmer Really Looks Like." Cartoon. *Country Gentleman*, 30 July 1921, 3.
Rome, Adam Ward. "American Farmers as Entrepreneurs, 1870–1900." *Agricultural History* 56, no. 1 (January 1982): 37–49.
Rosowski, Susan J. *The Voyage Perilous: Willa Cather's Romanticism*. Lincoln: University of Nebraska Press, 1986.
Rosowski, Susan J., and Charles W. Mignon. Preface to *O Pioneers!* by Willa Cather, edited by Rosowski and Mignon, vii–xi. Lincoln: University of Nebraska Press, 1992.
Rothstein, Morton. "Frank Norris and Popular Perceptions of the Market." *Agricultural History* 56, no. 1 (January 1982): 50–66.
Rouse, Blair. *Ellen Glasgow*. New York: Twayne, 1962.
Rowell, Charles H. "The Quarters: Ernest Gaines and the Sense of Place." *Southern Review* 21, no. 3 (Summer 1985): 733–50.
Royce, Edward. *The Origins of Southern Sharecropping*. Philadelphia: Temple University Press, 1993.
"Rural Opposition to Hog Farms Grows." *New York Times*, 22 September 1997, A18.
Sachs, Carolyn E. *Gendered Fields: Rural Women, Agriculture, and Environment*. Boulder, Colo.: Westview Press, 1996.
Salamon, Sonya. *Prairie Patrimony: Family, Farming, and Community in the Midwest*. Chapel Hill: University of North Carolina Press, 1992.
Sarver, Stephanie L. *Uneven Land: Nature and Agriculture in American Writing*. Lincoln: University of Nebraska Press, 1999.
Schaller, Neill. "Background and Status of the Low-Input Sustainable

Agriculture Program." In *Sustainable Agriculture Research and Education in the Field: A Proceedings*, 22–31. Board on Agriculture, National Research Council. Washington, D.C.: National Academy Press, 1991.

Schlebecker, John T. *Whereby We Thrive: A History of American Farming, 1607–1972*. Ames: Iowa State University Press, 1975.

Schmidt, Jan Zlotnik. "Ellen Glasgow's Heroic Legends: A Study of *Life and Gabriella*, *Barren Ground*, and *Vein of Iron*." *Tennessee Studies in Literature* 26 (1981): 117–41.

Schuck, Peter. "Tied to the Sugar Lands." *Saturday Review*, 6 May 1972, 36–42.

Schweninger, Loren. "A Vanishing Breed: Black Farm Owners in the South, 1651–1982." *Agricultural History* 63, no. 3 (Summer 1989): 41–60.

Shelton, Frank W. "Of Machines and Men: Pastoralism in Gaines's Fiction." In *Critical Reflections on the Fiction of Ernest J. Gaines*, edited by David C. Estes, 12–29. Athens: University of Georgia Press, 1994.

Sherman, Caroline B. "The Development of American Rural Fiction." *Agricultural History* 12, no. 1 (1938): 67–76.

Shillinglaw, Susan. "California Answers *The Grapes of Wrath*." In *John Steinbeck: The Years of Greatness, 1936–1939*, edited by Tetsumaro Hayashi, 145–64. Tuscaloosa: University of Alabama Press, 1993.

Shiva, Vandana. "Masculinization of the Motherland." In *Ecofeminism*, edited by Maria Mies and Vandana Shiva, 108–15. London: Zed Books, 1993.

———. *Monocultures of the Mind: Perspectives on Biodiversity and Biotechnology*. London: Zed Books, 1993.

Shrivastava, Paul. "Societal Contradictions and Industrial Crises." In *Learning from Disaster: Risk Management after Bhopal*, edited by Sheila Jasanoff, 248–67. Philadelphia: University of Pennsylvania Press, 1994.

Sims, Calvin. "Japan: Food Poisoning." *New York Times*, 7 July 2000, A9.

Singmaster, Elsie. "An Early Spring." In *Early Stories from the Land: Short-Story Fiction from American Rural Magazines, 1900–1925*, edited by Robert G. Hays, 205–12. Ames: Iowa State University Press, 1995.

Slicer, Deborah. "Toward an Ecofeminist Standpoint Theory: Bodies as Grounds." In *Ecofeminist Literary Criticism: Theory, Interpretation, Pedagogy*, edited by Greta Gaard and Patrick D. Murphy, 49–73. Urbana: University of Illinois Press, 1998.

Smiley, Jane. Interview with Suzanne Berne. *Belles Lettres* 7, no. 4 (Summer 1992): 36–38.

———. *Moo*. New York: Ivy Books, 1998.

———. *A Thousand Acres*. New York: Fawcett Columbine, 1992.

Smith, Henry Nash. *Virgin Land*. Cambridge: Harvard University Press, 1970.

Smith, Raymond C. "New Conditions Demand New Opportunities." In *Farmers in a Changing World: The Yearbook of Agriculture, 1940*, edited by Gove Hambidge, 810–26. Washington, D.C.: Government Printing Office, 1940.

Smith, Terence. "Carter Embargoes Technology for Soviets; Limits Fishing Privileges and Sale of Grain in Response to 'Aggression' in Afghanistan." *New York Times*, 5 January 1980, 1, 6.

Spencer, G. R. "What a Farmer Really Looks Like." Cartoon. *Country Gentleman*, 3 December 1921, 13.

Steinbeck, John. *The Grapes of Wrath*. New York: Penguin, 1992.
———. *Their Blood Is Strong*. Pamphlet. San Francisco: Simon J. Lubin Society, 1938.
———. *Working Days: The Journals of The Grapes of Wrath, 1938–1941*. Edited by Robert DeMott. New York: Viking, 1989.
Stilgoe, John R. *Common Landscape of America, 1580–1845*. New Haven: Yale University Press, 1982.
Stouck, David. "Explanatory Notes." In *O Pioneers!*, edited by Susan J. Rosowski and Charles W. Mignon, 325–46. Lincoln: University of Nebraska Press, 1992.
Stout, Janis P. *Strategies of Reticence: Silence and Meaning in the Works of Jane Austen, Willa Cather, Katherine Anne Porter, and Joan Didion*. Charlottesville: University Press of Virginia, 1990.
Strange, Marty. *Family Farming: A New Economic Vision*. Lincoln: University of Nebraska Press, 1988.
*The Struggle of the Mussel Slough Settlers for Their Homes!* Visalia, Calif.: Delta Printing Establishment, 1880.
Summerville, James. "Rural America: An Index." *North Dakota Quarterly* 53, no. 4 (Fall 1985): 25–34.
Sykes, C. H. "What a Farmer Really Looks Like." Cartoon. *Country Gentleman*, 16 July 1921, 5.
Takada, Kazunori. "Bad Milk Leaves 12,000 Sick in Japan." Yahoo! News. 6 July 2000. 7 July 2000. http://dailynews.yahoo.com/h/nm/20000706/wl/japan_poisoning_dc_4html.
Taylor, Frank J. "California's 'Grapes of Wrath.'" *Forum*, 19 November 1939, 232–38.
"Then and Now." *Visalia Weekly Delta*, 4 June 1880.
Thiébaux, Marcelle. *Ellen Glasgow*. New York: Frederick Ungar, 1982.
Thompson, Paul B. *The Ethics of Aid and Trade*. Cambridge: Cambridge University Press, 1992.
Tilden, Freeman. "What a Farmer Really Looks Like." *Country Gentleman*, 2 July 1921, 6–7, 29.
Tillman, James S. "The Transcendental Georgic in *Walden*." *ESQ* 21, no. 3 (1975): 137–41
"Tractor 'Co-ops' Get U.S. Blessing." *Newsweek*, 5 September 1938, 33–34.
Turner, Frederick. "Cultivating the American Garden." In *The Ecocriticism Reader: Landmarks in Literary Ecology*, edited by Cheryll Glotfelty and Harold Fromm, 40–51. Athens: University of Georgia Press, 1996.
Turner, Lou. "No Justice for Black Farmers." April 1999. 31 January 2000. *News and Letters*. http://www.newsandletters.org/4.99_bw.htm.
Tyson Foods, Inc. "Vertical Integration." *Annual Report*. 1998. 10 July 2000. http://www.tyson.com/investorrel/98annual/s-verticalintegration.asp.
Uhl, Michael, and Tod Ensign. *GI Guinea Pigs: How the Pentagon Exposed Our Troops to Dangers More Deadly Than War: Agent Orange and Atomic Radiation*. N.p.: Playboy Press Book, 1980.
Unger, Irwin. *These United States*. Vol. 2. Boston: Little, Brown, 1982.
Urgo, Joseph R. *Willa Cather and the Myth of American Migration*. Urbana: University of Illinois Press, 1995.

"U.S. Patent on New Genetic Technology Will Prevent Farmers from Saving Seed." Rural Advancement Foundation International. 11 March 1998. 15 February 1999. http://www.rafi.org/genotypes/980311seed.html.
Valdez, Luis. "El Teatro Campesino—Its Beginnings." In *The Chicanos: Mexican American Voices*, edited by Ed Ludwig and James Santibañez, 115–19. Baltimore: Penguin Books, 1971.
———. *Luis Valdez—Early Works: Actos, Bernabé and Pensamiento Serpentino*. Houston: Arte Público Press, 1990.
Van Dyke, Henry. "The Red River of the North." *Harper's New Monthly Magazine* 40, no. 340 (May 1880): 801–17.
Varro. *On Agriculture*. In *Cato and Varro on Agriculture*, translated by William Davis Hooper and edited by G. P. Goold, 160–529. Cambridge: Harvard University Press, 1993.
"Vietnamese Farmers." Photo. *Washington Post*, 15 May 1994, A1.
Virgil. *The Georgics*. Translation and introduction by L. P. Wilkinson. London: Penguin Books, 1982.
Vobejda, Barbara. "Agriculture No Longer Counts." *Washington Post*, 9 October 1993, A1, A13.
"Voice of the Farmer." Narrated by Mike Wallace. Transcript. *60 Minutes*. CBS. WYOU, Scranton, Pennsylvania, 9 April 2000.
Waldmeir, John C. "A New Source for Frank Norris's 'Epic of The Wheat.'" *English Language Notes* 31, no. 3 (March 1994): 53–59.
Wallace, Allison. E-mail to the author. 3 July 2000.
Walter, G., and A. Reisner. "Midwestern Land-Grant University Scientists' Definitions of Sustainable Agriculture: A Delphi Study." *American Journal of Alternative Agriculture* 9, no. 3 (1994): 109–21.
"We, Too, Are Manufacturers." *Farm Journal*, November 1890, 198.
White, Richard. "'Are You an Environmentalist or Do You Work for a Living?': Work and Nature." In *Uncommon Ground: Toward Reinventing Nature*, edited by William Cronon, 171–85. New York: W. W. Norton, 1995.
White, William Allen. "The Business of a Wheat Farm." *Scribner's Magazine* 22, no. 5 (November 1897): 531–48.
"Why Should You Care about Agriculture?" Pamphlet. Harrisburg: Pennsylvania Department of Agriculture, n.d.
"Why the Fish Are Dying." Editorial. *New York Times*, 22 September 1997, A26.
"Why Young Women Are Leaving Our Farms." *Literary Digest*, 2 October 1920, 56–58.
Wiebe, Robert H. *The Search for Order, 1877–1920*. New York: Hill and Wang, 1967.
Wilkinson, L. P. *The Georgics of Virgil: A Critical Survey*. London: Cambridge University Press, 1969.
Williams, Gaar. "What a Farmer Really Looks Like." Cartoon. *Country Gentleman*, 27 August 1921, 5.
Williams, Raymond. *The Country and the City*. New York: Oxford University Press, 1973.
Williams, Robert C. *Fordson, Farmall, and Poppin' Johnny: A History of the Farm Tractor and Its Impact on America*. Urbana: University of Illinois Press, 1987.
Wilson, Warren H. "Social Life in the Country." In *Country Life*, edited by J. P.

Lichtenberger, 119–30. Philadelphia: American Academy of Political and Social Science, 1912.

Wojcik, Jan. *The Arguments of Agriculture*. West Lafayette, Ind.: Purdue University Press, 1989.

Woodress, James. *Willa Cather: A Literary Life*. Lincoln: University of Nebraska Press, 1987.

Worster, Donald. "Thinking Like a River." In *The Wealth of Nature: Environmental History and the Ecological Imagination*, 123–34. Oxford: Oxford University Press, 1994.

Wyatt, David. *The Fall into Eden*. Cambridge: Cambridge University Press, 1990.

Wyckoff, Walter A. "The Workers: An Experiment in Reality: IV—A Farm-Hand." *Scribner's Magazine* 22, no. 5 (November 1897): 549–60.

Yamaguchi, Mari. "Bad Milk Scandal Worries Japanese." *Tribune* (Scranton, Pa.), 31 August 2000, C12.

Yarbro-Bejarano, Yvonne. "From *acto* to *mito*: A Critical Appraisal of the Teatro Campesino." In *Modern Chicano Writers*, edited by Joseph Sommers and Tomás Ybarra-Frausto, 176–85. Englewood Cliffs, N.J.: Prentice-Hall, 1979.

Zagarell, Sandra A. "Narrative of Community: The Identification of a Genre." *Signs* 13 (Spring 1988): 498–527.

# INDEX

Adams, Edward F., 48
AFL-CIO, 131
Agent Orange, 140, 206 (n. 8)
Agrarianism, Jeffersonian, 11–12, 15, 22, 39; *Barren Ground* and, 82–83, 92; Country Life movement and, 67; survey grid and, 160; *O Pioneers!* and, 66–67; *A Thousand Acres* and, 163; Wendell Berry and, 174. *See also* Agrarian myth
Agrarian myth, 4, 11–12, 36, 57, 136; farmers as male, 23, 65, 67, 82; new, 18–19
Agricultural ladder, 103–4
Agriculture: alternative, 22, 24, 170–72, 178–79, 208 (n. 9); landownership and, 43–44. *See also* Community, entrepreneurial; Community, yeoman; Farmers; Women, farm
Agriculture, industrial, 4, 5, 14–26, 60–61, 112, 158–60; agrarian radicalism and, 79; alternative agriculture and, 171–72; bonanza farms, 26–40; characteristics, 16, 61, 65–66, 83, 86, 135, 158, 163; cooperative marketing and, 45; Country Life movement and, 17–18, 67, 73–74, 84, 89–90; county extension agents and, 85; defenders of, 158–59, 170; defined, 201 (n. 1); education and, 38, 76–77; intelligence and, 37–38, 65, 68, 74, 75, 84; labor and, 37, 96, 98, 124–25, 129–32, 143, 158, 206 (n. 6); land-grant universities and, 169; as masculine, 66; nature and, 39–40, 44, 47, 48, 55, 61, 75, 84, 93; the pastoral and, 6, 9–10; pesticides and, 139–40; race and, 125, 134, 144–45; Smiley and, 162; in the South, 145, 147, 204 (n. 20); Steinbeck and, 99, 115; survey grid and, 160–61; tractors and, 102–7, 205 (n. 5); as warfare, 29, 141
Agriculture, preindustrial, 12–15, 16, 44, 66, 180; nature and, 48
*Agweek*, 205 (n. 1)
Allis-Chalmers, 104
American Farm Bureau (AFB), 22, 97, 116, 205 (n. 8)
American Farmland Trust, 3
Amish, 102, 125, 152; tractors and, 159–60
Animal rights, 158
Anthropology, 101
Archer-Daniels-Midland, 4, 170, 208 (n. 7)
Arthurdale, W.Va., 107
Arvin Sanitary Camp, 97
Associated Farmers, 99, 110, 119, 204 (nn. 4, 7); campaign against *The Grapes of Wrath*, 97–98, 111, 121–22
Association for the Study of Literature and the Environment, 9
*Atlantic Monthly*, 27, 28, 29, 68
Aztlán, 143

Baker, Ray Stannard, 27, 38, 40
Baldwin, C. B., 97
"Battle Hymn of the Republic," 123
Berry, Wendell, 24, 156, 172, 183; Earl Butz and, 174–76; "The Farm," 179–84; *Remembering*, 175–78; *A Timbered Choir*, 179; tractor-horse debate and, 207 (n. 18); *The Unsettling of America*, 174; Virgil and, 181, 182; "Writer and Region," 176
Beus, Curtis E., 171–72
Bhopal, India, 60
Bigelow, Poultney, 28–29, 33–34, 35–37, 40, 56, 201 (n. 3)
Biotechnology, 4, 170
Bishop, Holmes, 119

Black farmers, 22, 24; ages of, 149; extinction of, 144–46; numbers of, 22, 144; racial discrimination suit, 4, 23, 145, 149–50
Black Farmers and Agriculturalists Association, 146
Bonanza farms, 14, 22, 26–39, 104; maps and, 40, 50
Bond, Tonette, 83
Book farming. *See* Agriculture, industrial
Boone, Robert, 207 (n. 12)
Boren, Lyle, 122
Bovine spongiform encephalopathy. *See* Mad-cow disease
Bracero program, 125
Brown, John, 123
Bryan, William Jennings, 203 (n. 14)
Buell, Lawrence, 7
Burkett, Charles W., 64, 65
Butterfield, Kenyon, 17, 18–19, 73
Butz, Earl, 141, 169, 170, 172–74, 176; Wendell Berry and, 174–75

Cajuns, 23, 147
California, 40, 48, 100; farming in, 22, 97–100, 115, 124–25, 130–32; farm labor unions in, 128; Iowa and, 207 (n. 2)
*California Farmer*, 128, 129
*California Guide Book*, 43
California Rural Legal Assistance, 205 (n. 2)
California State Department of Public Health, 139
Camp, W. B., 121
Carden, Mary Paniccia, 7
Carlson, Ron, 164
Carter, Jimmy, 141
Cass, George W., 29
Cass-Cheney farm, 33
Cass farm, 27, 44
Casson, Herbert N., 68, 203 (n. 6)
Cather, Willa, 66, 202 (n. 1); "Nebraska: The End of the First Cycle," 67; *O Pioneers!*, 7, 22, 66–80, 83, 87, 92; as urban agrarian, 67–68
Cato, 201 (n. 3)
Chaffee, E. W., 33
Chavez, Cesar, 23, 125, 144, 205 (n. 3); and David-and-Goliath analogy, 130, 205 (n. 1); death, 127, 205 (n. 2); obituary, 128–30; pesticides and, 139; on race, 134, 206 (n. 4); Teamsters and, 131, 206 (n. 10)
Cherny, Robert W., 67
*Chicago Daily News*, 82
Civil War, 29, 123, 144, 146
Cloning, 48
Coffin, C. C., 28, 29, 36–37, 40, 56
Commission on the Conservation of Natural Resources, 203 (n. 7)
Community, entrepreneurial: defined, 101–2; *The Grapes of Wrath* and, 108–9, 112; *Of Human Kindness* and, 114, 117–19, 122; today, 125–26; tractor advertising and, 104
Community, yeoman: defined, 101–2; in *The Grapes of Wrath*, 108, 113–14; neighboring in, 101; today, 125
Community-sustained agriculture, 22
Computers, 25–26, 60
Conagra, 60
Confederación de Uniones de Obreros Mexicanos, 128
Conservation movement, 17, 75
*Country Gentleman*, 81, 82, 104, 105
Country Life Commission, 17, 64, 76, 203 (n. 7)
Country Life movement, 17–18, 67–68, 71, 204 (n. 22); *Barren Ground* and, 87–90; *O Pioneers!* and, 75–77
Creoles, 147, 207 (n. 14)
*Cultivator and Country Gentleman*, 14, 27, 28

Dalrymple, Oliver, 33
Dante, 178
Davenport, Eugene, 203 (n. 6)
Day, Mark, 206 (n. 4)
Debt, farm, 88, 170
Delano grape strike, 127, 130–34, 140. *See also* Chavez, Cesar

DeMarco, Susan, 207 (n. 1)
Demonstration farms, 77
DiGiorgio Fruit Corporation, 130, 131
Dillon, Charles, 68
Dorr, Tom, 26, 28, 45, 60
Dow Chemical Corporation, 140
Dunlap, Riley E., 171–72
Dunne, John Gregory, 134
Dust Bowl, 21, 115, 124

"An Early Spring" (Singmaster), 81
Ecocriticism, 9
Equal Rights Amendment, 22

Farmall tractors, 102–7, 205 (n. 5)
Farm Crisis (1980s), 10, 170, 207 (n. 2)
Farmers: grain prices and, 39, 169; as male, 11, 65, 81–82; numbers of, 3, 22, 96, 125–26; types of, 17, 18, 73–74, 81–82, 84, 203 (n. 6). *See also* Agriculture; Agriculture, industrial; Agriculture, preindustrial; Black farmers; Women, farm
Farmers Home Administration, 22, 145
*Farm Journal*, 17
Farmland, 3; value of, 170
Farm novel: defined, 19–21
Farms: consolidation of, 103; geometry of, 109; numbers of, 125–26; ownership of, 43–44; size of, 96, 125
Farm Security Administration (FSA), 97, 107
Federal Reserve, 170
Fiedler, Leslie, 6
Fink, Deborah, 79
Fisher, Ann H., 164
Fite, Gilbert, 145, 204 (n. 20)
Flanagan, John T., 19
Food, as a weapon, 141
Food and Farming in American Life and Letters Symposium, 9
Foot-and-mouth disease, 208 (n. 8)
Ford, John, 103
Forty Acres, 144
Foster, Murphy, 152
French, Warren, 204 (n. 1)
Fuller, Wayne E., 201 (n. 6)

Gaines, Ernest, 23, 143, 149, 155; *A Gathering of Old Men*, 125, 143–56; pastoralism and, 143–44
Gallo, Julio, 130; obituary, 128–29
Garden of Eden, 7, 11
Garland, Hamlin, 80
George, Henry, 43
Georgic, 8–11, 24, 144, 179
Gerlach, Dwayne, 25–26, 28, 60
Gilman, Charlotte Perkins, 64
Glasgow, Ellen, 204 (nn. 18, 23); *Barren Ground*, 22, 65, 80, 82–92; reviews of *Barren Ground* and, 83
Global positioning systems, 4, 26, 60
Godbold, E. Stanly, 204 (nn. 18, 24)
Golden Age, 6, 8
*Good Housekeeping*, 64
Grami, William, 206 (n. 10)
Grandin Farm, 34
Grant, Gary, 146
*The Grapes of Wrath* (film), 103
Great Britain, 171, 208 (n. 8)
Great Depression, 21, 23, 96, 103
Greeley, Horace, 43, 44, 57
Green Revolution, 138, 139
Grid survey. *See* Survey grid system
Gustafson, Neil, 73–74

Hardin, Clifford Morris, 172
Harper's Ferry, Va., 123
*Harper's New Monthly Magazine*, 27, 28, 29
Harrison, Elizabeth, 82
Harwood, William Sumner, 18
Hayes, Rutherford B., 27
Hernández, Guillermo, 135
Hesiod, 201 (n. 3)
Hightower, Jim, 207 (n. 1)
*Hoard's Dairyman*, 105
Hollister, 41
Holmes, Roy Hinman, 68
Homestead Act, 12, 29, 43, 147
Hoover, Herbert, 114

Hooverville, 113
*Hudson Review*, 176, 180

*I'll Take My Stand*, 92
*Independent*, 68
*Indianapolis News*, 82
Interchurch World Movement, 89
International Harvester Company (IH), 102, 104, 116; monopoly and, 107
Iowa, 25–26, 45, 161, 164, 207 (n. 2)
Irrigation, 39, 48–49
Itliong, Larry, 142

Jack-in-the-Box, 170
Jackson, Andrew, 43
Jefferson, Thomas, 11–12, 67, 178; grid survey and, 50, 160; *Notes on the State of Virginia*, 11, 162–63. *See also* Agrarianism, Jeffersonian
Jewett, Sarah Orne: *The Country of the Pointed Firs*, 176
Jim Crow laws, 144, 148
Julian, George, 43

Kentucky, 174
Kern County, Calif., 97, 98, 121
Kern County Agricultural Commission, 140
*King Lear*, 5, 162, 168
Kolodny, Annette, 6

Lancaster County Cooperative Extension Service, 157
*Lancaster Farming*, 157, 159
Land Ordinance (1785), 160
Lawrence, D. H., 6
Leopold, Aldo, 179, 183; land ethic and, 178–79
Le Sueur, Meridel, 169
*Library Journal*, 164
Light, Frank, 129–30
Lisca, Peter, 99
Little Rock, Ark., 149
Lodgson, Gene, 207 (n. 18)
Louisiana, 143, 146, 147, 154

*McClure's*, 27, 40, 67, 68
McCormick, Cyrus, 33
McDowell, Frederick, 91, 204 (n. 21)
MacGarr, Llewellyn, 202 (n. 5)
McGee, Leo, 207 (n. 12)
McWilliams, Carey, 101, 124–25, 204 (n. 4), 205 (n. 5)
Mad-cow disease, 4, 47, 170–71
*El Malcriado*, 206 (n. 10)
Manchester College, 174
"The Man from the City" (Hay), 80–81
"The Man with a Hoe" (Markham), 54
*The Man with a Hoe* (Millet), 54
Maps. *See* Bonanza farms: maps and; Survey grid system
Markham, Edwin, 54
Marx, John Walter, 205 (n. 6)
Marx, Leo, 6
Matthiessen, Peter, 130, 132
Mennonites, 35
Meyer, Roy, 19, 20
Mignon, Charles, 202 (n. 1)
Millet, François, 54
Milton, John, 178
Mitchell, Ruth Comfort, 23, 96, 125, 204 (n. 4); at Associated Farmers convention, 97–98, 119; *Of Human Kindness*, 23, 96–97, 100–101, 109–12, 114–25, 133
Monocultures, 158, 175–76, 180
Monopoly, 167
Monsanto, 60
Morley, C. Seldon, 140
Morrison, Toni: *The Bluest Eye*, 146
Mossback. *See* Farmers: types of
Muir, John, 75
Murphy, John, 7
Mussel Slough, 201 (n. 5), 202 (n. 7); shootout at, 42–43, 48–49

Narrative of community, 176
National Farmworkers Union. *See* United Farm Workers Union
Nebraska, 7, 67, 78
New Agriculture. *See* Agriculture: industrial
*New Farm*, 174

New farmer. *See* Agriculture, industrial; Farmers: types of
Newswanger, Everett, 157, 158, 159
*Newsweek*, 107
*New York Times*, 139, 206 (n. 10)
*New York Times Book Review*, 164
New York University, 174
Nixon, Richard, 141, 169, 172
Norris, Frank, 61; knowledge of agriculture, 40–41; maps and, 50–51; Mussel Slough and, 49; *The Octopus*, 22, 27, 39–61, 168; *O Pioneers!* and, 66, 74; *The Pit*, 61; view of women, 56; *The Wolf*, 61
North Dakota, bonanza farms and, 14, 22, 26–29
Northern Pacific Railroad, 27, 36
North Manchester, Ind., 174

*Oklahoma City Times*, 122
Old farmer. *See* Agriculture, preindustrial; Farmers: types of
Olestra, 4
*Omaha World-Herald*, 82
*Orion*, 174
*Outlook*, 68

Pastoral, 10–11, 48, 82, 93, 143; in American literature, 6–7, 9
Penn State University, 159
Pesticides, 131, 139–40, 158, 206 (n. 6); Bhopal disaster and, 60
Philadelphia, 161
Pinchot, Gifford, 75
Pizer, Donald, 42
Populism, 34, 45, 68, 202 (n. 4), 203 (nn. 14, 15); farm reformers' fear of, 79, 204 (n. 22); producerism and, 44
Port William, Ky., 176
Power, James B., 27, 28, 33, 36, 40
Progressive Era, 67, 75
Purdue University, 172

Ralston Purina, 172
Randall, John H., 7
La Raza, 134
Red River Valley, 26, 27, 28, 29; machinery in, 33, 37
*Review of Reviews*, 68
Reynolds, Guy, 67
Rockefeller, John D., 89
Roosevelt, Theodore, 17, 64, 67, 75
Rosowski, Susan, 7, 202 (n. 1)
Rouse, Blaire, 204 (nn. 18, 21)

Salamon, Sonya, 101
San Anita Rancho, 41
San Joaquin Valley, 23, 127, 134; irrigation in, 48–49
San Luis, Ariz., 127
*Saturday Review*, 152
Schaller, Neill, 208 (n. 9)
Schenley Industries, 131
Schmidt, Jan Zlotnik, 82
*Scranton Tribune*, 95
*Scribner's Magazine*, 14, 27
Shakespeare, 162
Sharecropping, 103, 144, 147, 148, 207 (n. 13)
Shelton, Frank W., 143
Sheppard, Mildred, 95, 96
Shillinglaw, Susan, 204 (n. 1)
Shiva, Vandana, 175
*60 Minutes*, 22
Slavery, 21, 23, 123, 148; slave quarters and, 146–47
Smiley, Jane, 5, 156, 179; farming and, 162; *Moo*, 169, 170, 178; *A Thousand Acres*, 5, 24, 160–69, 175, 177, 178, 183
Smith, Henry Nash, 6
Smith-Lever Act, 71–72, 87, 203 (n. 9)
Southern Pacific Railroad, 43, 206 (n. 4)
Stanford University, 174
Steinbeck, John, 97, 108, 125; Associated Farmers and, 99; "Battle Hymn of the Republic" and, 123; campaign against *The Grapes of Wrath* and, 98, 111, 121–22; community and, 99–100, 117; *The Grapes of Wrath*, 23, 96, 104, 107–10, 112–14, 122,

123, 124, 133; industrial agriculture and, 99, 100, 115
Stone Mountain, Ga., 95
Stout, Janis, 7
Strange, Marty, 201 (n. 1)
Sugar bowl (La.), 146
Survey grid system, 50, 168, 178; characteristics, 161, 165, 167; land distribution and, 160–61
Swain, Francis, 26

Taylor, Frank J., 205 (n. 7)
Teamsters, 131, 137, 206 (n. 10)
Teatro Campesino (Farmworkers Theater), 131, 132–34, 137, 143, 206 (n. 5). *See also* Valdez, Luis
Technology, 20, 45, 65, 134, 158; restrained use of, 9, 102, 107, 159–60, 171
Tractors, 26, 102–7, 145, 204 (n. 20), 205 (nn. 5, 6); Amish and, 159–60; and redesigning of fields, 109; horse-tractor debate, 152, 207 (n. 18); migrant farmworkers and, 116; as status symbol, 87, 103
Tyson Foods, 61

United Farm Workers Organizing Committee. *See* United Farm Workers Union
United Farm Workers Union (UFW), 127, 128, 132, 134, 139–40; Teamsters and, 131, 137
University of Nebraska, 203 (n. 8)
Urban agrarians, 17–18, 20, 66–67, 71–72, 84; Cather as, 67–68; conservation and, 75. *See also* Country Life movement
U.S. Census Bureau, 126
U.S. Civil Rights Commission, 23, 144
U.S. Department of Agriculture (USDA), 38, 207 (nn. 1, 18); industrial agriculture and, 66, 84, 89; migrant farmworkers and, 98; 1940 *Yearbook*, 102; racial discrimination case and, 4, 23, 145, 149–50; tractors and, 103

Valdez, Luis, 23, 125, 155, 205 (n. 3); *actos* and, 132–34, 151, 155; *Las Dos Caras del Patroncito*, 134–37; *Vietnam Campesino*, 137–43
Van Dyke, Henry, 28, 29, 35, 40
Varro, 201 (n. 3)
Vietnam, 63, 143
Vietnam War, 137, 138, 139, 142; Agent Orange and, 140, 206 (n. 8)
Virgil, 8, 12, 29, 181, 182; *Aeneid*, 7; *Eclogues*, 7, 66; *Georgics*, 7–8, 141, 180
*Visalia Weekly Delta*, 202 (n. 8)
Voting Rights Act (1965), 22

Wallace, Henry C., 205 (n. 8)
Washington, D.C., 50, 161
*Washington Post*, 22, 63, 65
Washington State, 170
"What a Farmer Really Looks Like," 81–82
White, William Allen, 14, 28, 29–33, 34–35, 37
Williams, Gaar, 82
Williams, Robert C., 116
Williamsport, Pa., 15
Wilson, Warren H., 86
Women, farm, 21, 22, 23; Country Life movement and, 64; invisibility of, 63–65, 66; nature and, 93, 162; separate spheres and, 37, 71–72, 87; work and, 64, 65
*World's Work*, 68
World War I, 23, 144, 149, 204 (nn. 18, 22)
World War II, 96, 125, 140, 152
Worldwatch, 3
Wyatt, David, 202 (n. 11)
Wyckoff, Walter A., 14, 15

Yeoman, 8; defined, 65; Jefferson's, 11–12, 18, 42, 66. *See also* Agrarian myth
Young, Sanborn, 98